WITHDRAWN
UTSA LIBRARIES

Media and the Politics of Arctic Climate Change

Also by Miyase Christensen

CONNECTING EUROPE ONLINE TERRITORIES: GLOBALIZATION, MEDIATED PRACTICE AND SOCIAL SPACE (*co-edited with André Jansson and Christian Christensen*)

MEDIA, SURVEILLANCE AND IDENTITY: SOCIAL PERSPECTIVES (*co-edited with André Jansson*)

SHIFTING LANDSCAPES: Film and Media in European Context (*co-edited with Nezih Erdogan*)

Also by Annika E. Nilsson

ULTRAVIOLET REFLECTIONS: Life under a Thinning Ozone Layer

GREENHOUSE EARTH

Also by Nina Wormbs

THE SCIENCE–INDUSTRY NEXUS: History, Policy, Implications (*co-edited with Karl Grandin and Sven Widmalm*)

Media and the Politics of Arctic Climate Change
When the Ice Breaks

Edited by

Miyase Christensen
Stockholm University and KTH Royal Institute of Technology, Sweden

Annika E. Nilsson
Stockholm Environment Institute, Sweden

and

Nina Wormbs
KTH Royal Institute of Technology, Sweden

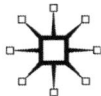

Introduction, selection and editorial matter © Miyase Christensen, Annika E. Nilsson and Nina Wormbs 2013

Individual chapters © Contributors 2013

All rights reserved. No reproduction, copy or transmission of this publication may be made without written permission.

No portion of this publication may be reproduced, copied or transmitted save with written permission or in accordance with the provisions of the Copyright, Designs and Patents Act 1988, or under the terms of any licence permitting limited copying issued by the Copyright Licensing Agency, Saffron House, 6–10 Kirby Street, London EC1N 8TS.

Any person who does any unauthorized act in relation to this publication may be liable to criminal prosecution and civil claims for damages.

The authors have asserted their rights to be identified as the authors of this work in accordance with the Copyright, Designs and Patents Act 1988.

First published 2013 by
PALGRAVE MACMILLAN

Palgrave Macmillan in the UK is an imprint of Macmillan Publishers Limited, registered in England, company number 785998, of Houndmills, Basingstoke, Hampshire RG21 6XS.

Palgrave Macmillan in the US is a division of St Martin's Press LLC, 175 Fifth Avenue, New York, NY 10010.

Palgrave Macmillan is the global academic imprint of the above companies and has companies and representatives throughout the world.

Palgrave® and Macmillan® are registered trademarks in the United States, the United Kingdom, Europe and other countries

ISBN: 978–1–137–26622–4

This book is printed on paper suitable for recycling and made from fully managed and sustained forest sources. Logging, pulping and manufacturing processes are expected to conform to the environmental regulations of the country of origin.

A catalogue record for this book is available from the British Library.

A catalog record for this book is available from the Library of Congress.

Library
University of Texas
at San Antonio

For our children Lara, Dane, Vendela, Malva and Katja

Contents

List of Illustrations	viii
Preface and Acknowledgments	x
Notes on Contributors	xii

1. Globalization, Climate Change and the Media: An Introduction — 1
   Miyase Christensen, Annika E. Nilsson and Nina Wormbs

2. Arctic Climate Change and the Media: The News Story That *Was* — 26
   Miyase Christensen

3. Eyes on the Ice: Satellite Remote Sensing and the Narratives of Visualized Data — 52
   Nina Wormbs

4. An Ice-Free Arctic Sea? The Science of Sea Ice and Its Interests — 70
   Sverker Sörlin and Julia Lajus

5. Signals from a Noisy Region — 93
   Annika E. Nilsson and Ralf Döscher

6. A Question of Scale: Local versus Pan-Arctic Impacts from Sea-Ice Change — 114
   Henry P. Huntington

7. Under the Ice: Exploring the Arctic's Energy Resources, 1898–1985 — 128
   Dag Avango and Per Högselius

8. Changing Arctic – Changing World — 157
   Miyase Christensen, Annika E. Nilsson and Nina Wormbs

Index — 173

List of Illustrations

Figures

1.1	Sea-ice extent 1979–2007	3
3.1	Arctic sea ice 'melts to all-time low'	60
3.2	Satellites witness lowest Arctic ice coverage in history	62
3.3	Left, original AS17-148-22727 from Apollo 17 in 1972. Right, NASA Blue Marble from 2002	65
7.1	Coal-storage area at the Svea mine, Spitsbergen, in the early 1920s	135
7.2	Pier and loading facilities at Longyear City, Spitsbergen, early 1910s	136
7.4	Sketch of the 'fender platform', developed in the early 1980s by Norwegian and Canadian companies	149
7.5	Built to cut through thick sea ice en route to drill sites and to withstand moving pack ice during drilling operations	150

Maps

6.1	Map of the Arctic	118
7.3	Map on the front page of a Swedish coal-mining prospectus from 1916, presenting the Spitsbergen energy resources as easily accessible, ready to plug into the larger infrasystems of Scandinavia	139

Tables

2.1	News pieces published 2003–2006	35
2.2	News pieces published 2007–2010	35
2.3	News pieces published 2003–2010	36
2.4	Sectional breakdown of the articles containing climate change, the Arctic, and sea ice published in *The Guardian* in the period 2003–2010	37
2.5	Sectional breakdown of the articles containing climate change, the Arctic, and sea ice published in *The New York Times* in the period 2003–2010	37

2.6 Sectional breakdown of the articles containing climate change, the Arctic, and sea ice published in *Dagens Nyheter* in the period 2003–2010   37
2.7 Frames, subframes and topics   46

Preface and Acknowledgments

The Arctic has fascinated us for centuries. Through the stories of scientists, explorers, novelists and filmmakers, this polar region has been as much a part of popular culture and imagination as it has of specialist discourses. In Philip Pullman's *Northern Lights* (aka *The Golden Compass*), the heroine Lyra rides an armored polar bear named Iorek Byrnison with 'the Aurora...swaying above them in golden arcs and loops and all around...the bitter arctic cold and the immense silence of the North' (Pullman, 1995, *Northern Lights* [New York: Scholastic], pp. 182–3). Through Pullman's *His Dark Materials* trilogy, translated into numerous languages and adapted into film, yet another generation has encountered the frozen North. In his story the Arctic is not only the center of the world, but of all worlds.

As researchers who joined forces for the intellectual exercise of an Arctic excursion, we wish to extend our thanks to our funding agency, the Swedish Research Council FORMAS, for generously rewarding our research project, 'Models, Media and Arctic Climate Change'. This book, which brings together popular, scientific and historical accounts, was conceived and written during the course of this project. As participants, we would like to thank our project leader, Sverker Sörlin, and fellow partner Ralf Döscher for their ideas and work.

As editors of this volume, we would like to gratefully acknowledge, first and foremost, the contributions of our authors. We are sincerely indebted to all our reviewers who have provided invaluable insights and stimulating discussions at various stages along the way: Patrick Burkart, Christian Christensen, Monika Djerf-Pierre, Paul Edwards, James Fleming, Sabine Höhler, Paul Josephson, Igor Krupnik and Peder Roberts.

We are grateful to Susanna Lidström, for her invaluable work and assistance towards the final stages of the project.We are also indebted to our institutions, the Stockholm Environment Institute and the Division of History of Science, Technology and Environment, KTH Royal Institute of Technology, for providing material support and intellectual resources, and to our colleagues for blessing us with their cheerful and friendly presence while we were putting together this volume.

On the personal level, we wish to offer our sincerest thanks to our families and friends for simply being there for us and for enduring the piles of drafts and late-night reviewing.

We hope that this collection offers a glimpse of the worldly and scholarly challenges that lie ahead of us in our Arctic futures.

Miyase Christensen
Annika E. Nilsson
Nina Wormbs
Stockholm, February 2013

Notes on Contributors

Dag Avango is a researcher in the Division of History of Science, Technology and Environment at KTH Royal Institute of Technology, Stockholm, Sweden. His primary research interest is polar history, both Arctic and Antarctic, with a focus on the long-term historical development of large-scale natural resource extraction and its relation to scientific research and geopolitics. He is the author/co-author of numerous articles and books.

Miyase Christensen is Professor of Media and Communication Studies at Stockholm University and Guest Professor at the Division of History of Science, Technology and Environment at KTH Royal Institute of Technology in Stockholm, Sweden. She is the editor of *Popular Communication: International Journal of Media and Culture* and the Chair of the Ethnicity and Race in Communication Division of ICA (International Communication Association). Her most recent books as co-editor include *Online Territories: Globalization, Mediated Practice and Social Space* (2011), *Media, Surveillance and Identity* (2013) and *Understanding Media and Culture in Turkey: Structures, Spaces, Voices* (in press). Her research focuses, from a social-theory perspective, on globalization–transnationalization processes and social change; technology, culture and identity; and politics of popular communication. She is currently conducting funded projects including 'Cosmopolitanism from the Margins: Mediations of Expressivity, Social Space and Cultural Citizenship.'

Ralf Döscher is a senior researcher at Rossby Centre, the climate-modeling unit at the Swedish Meteorological and Hydrological Institute (SMHI). He holds a PhD in physical oceanography from the University of Kiel. He has more than 20 years of scientific experience in numerical studies of large-scale ocean circulation, marginal sea changes and Arctic climate change. His current field of interest covers changes in the Arctic atmosphere, sea ice, ocean and land, including interaction and feedback processes. He has led several EU projects and was co-coordinator of EU project DAMOCLES. He was responsible for the coordination of Arctic climate modeling of European and US activities in the EU project S4D. Currently, Ralf heads the Swedish national project ADSIMNOR on advancing Arctic climate simulation and process understanding. He has published extensively in his areas of research.

Notes on Contributors xiii

Per Högselius is Associate Professor in the Division of History of Science, Technology and Environment at KTH Royal Institute of Technology in Stockholm, Sweden. He holds an MSc in Engineering Physics and History of Technology from KTH, a PhD in Innovation Studies from Lund University and a Docent (habilitation) degree in History of Science and Technology from KTH. He has held fellowships at Bocconi University, Milan, and the Netherlands Institute of Advanced Study (NIAS). He has also worked as an independent expert for the OECD, PHARE and the Council of the Baltic Sea States. His research has focused on international relations in the history of science, technology and environment, with an emphasis on telecommunications, nuclear energy, electricity and natural gas – resulting, among other things, in a number of books published by leading academic publishers in Sweden, Germany, Britain and the United States. In Sweden, he is also active as an author of popular history books and technology- and culture-related newspaper essays.

Henry P. Huntington studies human–environment interactions in the Arctic, primarily among indigenous peoples. He earned his PhD in polar studies at the University of Cambridge, and has been involved in several sea-ice research projects in Alaska, Canada and Greenland. He is the author or co-author of more than three dozen scientific papers, as well as several books. These books include *Wildlife Management and Subsistence Hunting in Alaska* (1992) and, with Anne Salomon and Nick Tanape Sr., *Imam Cimiucia: Our Changing Sea* (2011). He is currently the Arctic Science Director for the Pew Charitable Trusts.

Julia Lajus is Associate Professor at the National Research University Higher School of Economics and at the European University, both in Saint Petersburg, where she is also Director of the Centre for the History of Science, Technology and Environment. She has a background in Arctic marine biology and holds a PhD in history from the Institute for the History of Science and Technology, Russian Academy of Sciences. She has published widely on the history of Soviet and Russian Arctic science and environment and has considerable experience in collaborative projects with European and American scholars, such as the 'History of Marine Animal Populations' program and ESF EUROCORES 'Boreas: Histories from the North'; the latter, as well as the current 'Assessing Arctic Futures' project, with Sverker Sörlin.

Annika E. Nilsson is Senior Research Fellow at Stockholm Environment Institute, Stockholm, Sweden. She has a PhD in environmental science and over 20 years of professional experience as a science writer. Her research is about communication at the science–policy interface, with focus on Arctic environmental change and its implications for society, including international politics. She has participated in several assessments about the Arctic as a science writer (focusing on pollution and on human development) and both planned and led the first phase of the Arctic Resilience Report as an Arctic Council project. Annika has published several popular science books on environmental change. She has been appointed a member of the Environmental Advisory Council by the Swedish government and is also a member of the Advisory Council to the Swedish Polar Research Secretariat.

Sverker Sörlin is Professor of the History of Ideas, Division of the History of Science, Technology and Environment at KTH Royal Institute of Technology, Stockholm, Sweden. He was the first director of the Swedish Institute for Studies in Education and Research, SISTER (2000–2003), and he has held visiting positions at the University of California, Berkeley (1993), the University of Cambridge (2004–2005), and the University of Oslo (2006). He has conducted research evaluations and public inquiries for governments and research councils. He has served on the boards of the universities in Umeå and Gothenburg, of public agencies and research foundations, and has had commissions at UNESCO and the European Science Foundation. From 1994 through 1998, and again from 2005 through 2009, he served on the Swedish government's Research Advisory Board. From 2006 to 2009 he served as President of the Swedish Committee for the International Polar Year 2007/08. *Narrating the Arctic* (with M. T. Bravo) was published in the United States in 2002. His most recent book in English is *Nature's End: History and the Environment* (with Paul Warde, 2009). Forthcoming is *Northscapes: History, Technology, and the Making of Northern Environments*, co-edited with Dolly Jörgensen.

Nina Wormbs is Associate Professor in the History of Science and Technology at the Division of the History of Science, Technology and Environment at KTH Royal Institute of Technology in Stockholm, Sweden. She has primarily written on conflicts around satellite and media technology and on allocation processes for radio frequencies. Her publications include *Vem älskade Tele-X? Konflikter om satelliter i Norden 1974–1989* (2003), *The Science–Industry Nexus: History, Policy, Implications* (2004, edited with Karl Grandin and Sven Widmalm), and *Radio och TV efter monopolet: en kamp om politik, pengar, publik och teknik* (2007, with Lars-Åke Engblom).

1
Globalization, Climate Change and the Media: An Introduction

Miyase Christensen, Annika E. Nilsson and Nina Wormbs

An unusual voyage

In the summer of 2011, the tanker *STI Heritage* left Houston, Texas and made the long, arduous journey to Thailand, eventually arriving with over 60,000 tons of condensed gas.[1] What made this trip special was not the start and end points (these are two major ports), but rather how, and how fast, the tanker made the journey. Instead of the traditional route via the Suez Canal, the *STI Heritage* picked up the condensed gas in Murmansk, Russia, and continued its journey towards Thailand via the Northeast Passage, a shipping lane running from Murmansk, along Siberia, ending at the Bering Strait. The use of this lane is, in and of itself, not unique, as historically portions of it have been navigable for two summer months each year. What made the *STI Heritage* voyage special, however, was the speed with which the vessel completed the entire route: eight days.[2] This was a record for the Northeast Passage (broken weeks later by a gas tanker that made the trip in just over six days), which has seen a dramatic reduction of summer ice over the past decade, making commercial use of the lane economically viable, at least if one extrapolates from the numbers. In 2009, only two commercial vessels made the voyage. In 2011, that number had increased to 18.[3]

Texas and Thailand. When discussing the Arctic, these two regions of the world are likely not the first two to come to mind. Yet the case of the *STI Heritage* is a stark illustration of the degree to which a specific environmental question – the melting of Arctic sea ice – has been transformed from an issue of local concern in a region of the world that has been relatively neglected in media terms, to an issue with not only local and regional but also global implications. The case of the *STI Heritage* journey from Texas to Thailand also crystallizes how the reduction of sea ice in the Arctic region has been the catalyst for a discussion that goes far beyond the environmental. When the shrinking ice creates possibilities to reduce use of the expensive Suez Canal and to avoid the perilous coast of eastern Africa, it also constitutes a signal of ongoing geopolitical changes connected to global resource demands and trade patterns.

The satellite images from the sea-ice minimum of 2007 also marked the starting point for a shift in the regional political climate in the Arctic, where Arctic policies focus increasingly on security and sovereignty (Huebert et al., 2012). The new political climate has foregrounded not only the efficacy of political governance of climate change, but also scenarios of a future ice-free Arctic as a well-publicized media image of climate change. In August 2012, the Arctic sea-ice extent was on its way to reach an even more dramatic level, spurring yet another burst in media attention as the new satellite data from the National Snow and Ice Data Center were released to the press. The questions that emerged from the press and media stories about the declining ice concern not only the ice itself. The race for natural resources, new shipping possibilities and the risk for political and commercial conflicts over the Arctic are also among the issues that gained significance. These stories started to build momentum in the aftermath of 2007 making the sea-ice minimum a media story that is relevant far beyond the Arctic and far beyond global climate science and policy. Thus, as a moment with broad scientific and political consequences, the 2007 ice minimum warrants close attention, and a multivalent approach, in order to pinpoint key issues at the intersection of science, politics and the media in the overall study of global climate change. As will be discussed shortly, in media studies the concept of mediatization refers to the tightly interconnected nature of social processes and the media. With an effort to illuminate the intricate role the media play in both the representation of climate change and in public understanding of scientific and political questions, we will utilize the frameworks of mediatization and media events (and eventization) as conceptual instruments.

This introductory chapter sets the stage for six substantive chapters and a concluding chapter that approach the 2007 sea-ice minimum from a range of different disciplines. Together they convey local, regional and global perspectives as well as considerations from the points of natural sciences, history, anthropology, science and technology studies, and media studies. We seek to highlight, in particular, the increasing role of the media in framing the present-day Arctic and its future, and of climate change. Media in this context are more than the coverage in newspapers, television and other venues. We place the sea-ice minimum into an analytical framework that highlights the increasing role of mediatization (simply put, media saturation) as a prominent social trend in our late-modern societies. Mediatization is enmeshed with globalization, commercialization and individualization (cf. Krotz, 2007), making the media dimension more complex and worthy of interdisciplinary scrutiny. In what follows we discuss concepts such as 'media events' and 'mediatization' in relation to the broader theoretical and conceptual tropes of seeing the environment as a social construct and environmental politics as intrinsically intertwined with other areas of international politics.

The sea-ice minimum as a media event

The new commercial possibilities that have been highlighted in the wake of the declining sea ice are but one aspect through which the Arctic has become a unique showcase for global climate change over the past few years. The other ways through which the region gained visibility include mediated images of polar bears on small remnants of sea ice and dramatic footage of calving glaciers where the Greenland ice sheet is moving into the sea chunk by chunk – images that are used by scientists and environmental organizations in their efforts to raise awareness about the consequences of climate change. In the late summer of

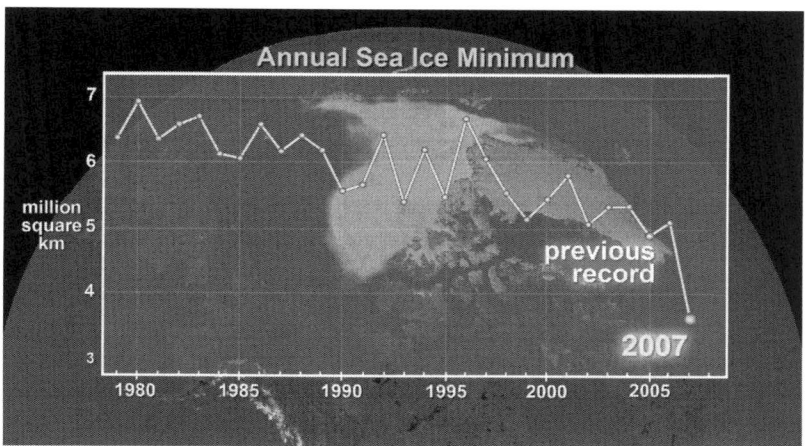

Figure 1.1 Sea-ice extent 1979–2007
Source: NASA.

2007, at the time of the 'sea-ice minimum', satellite technology played a key role in providing dramatic and influential images of the impacts of global warming in the Arctic when the data showed that the extent of sea ice in the Arctic Ocean had reached 23 percent below previously recorded low levels (2005) and 39 percent below the average over the period 1979–2000.[4] Moreover, the sea-ice decline well surpassed what most climate experts had anticipated. As such, the images based on the satellite data provided a dramatic preview of a post–global warming Arctic geography that quickly intensified debate on and around global climate change, turning it into a media *meta*-event.

In making the assertion that the sea-ice minimum was a 'media event', we construe 'media event' in the broader sense of a phenomenon, a moment, a 'happening' that draws significant media attention to the level of occupying considerable space in various forms and outlets of media. The concept was originally developed by Katz (1980) and Dayan and Katz (1992) in their bringing together both social scientific mass communication and cultural studies traditions to define the ways in which broadcast news covered certain instances and phenomena (cf. Couldry, Hepp and Krotz, 2009). In 1980, Katz wrote,

> The paradigmatic media event is one organized outside the media but which may well be transformed in the process of transmission. ... The element of high drama or high ritual is essential: the process must be emotion-laden or symbol-laden, and the outcome be rife with consequence. (p. 3)

Dayan and Katz's approach is one that looks upon media events as occasions 'where television makes possible an extraordinary shared experience of watching events at society's "centre"', (Couldry, 2003, p.61). Fiske (1994) takes media events as discursive events (not merely intense discourse produced based on an occasion) and draws attention to the importance of transborder flows and the need to understand media events in relation to the level of their transborder character. There are many examples of such media events in politics, ranging from the assassinations of John F. Kennedy and Martin Luther King in the 1960s to the more recent meta-event of the Arab Spring (see Christensen, 2013). With the aim of redeveloping the concept of media event in a global age, Couldry, Hepp and Krotz (2009, p. 9) argue that we need to understand media events not as products and mirrors of national cultures but of a highly complex global realm. The global realm is traversed by multiple nodes of connectivity (such as data and information flows) that influence both societal institutions (including media and politics) and the subjective domain of the public(s) and individuals. Put simply, in a global context both the production and reception of meaning/sense-making practices (such as media events) become more dispersed and complex. This necessitates correspondingly novel and multidimensional conceptual tools to capture both their character and due consequences.

Public debate about global climate change has also been closely linked with certain events, including the 1988 drought in North America and James Hansen's testimony to Congress the same year that linked together scientific and popular understandings of climate change and set the stage for moving international climate politics forward at the time. More recent examples are Hurricane Katrina in 2005 and the 2003 heat wave in Europe that killed almost 15,000 in France alone. As events that led to extensive media coverage of a range of issues on a global scale, they made apparent that adaptation to climate change was an issue of societal significance and would be a challenge also in rich countries with well-developed institutions.

While unexpectedness and immediacy (à la 'breaking news') are characteristic of certain events such as Hurricane Katrina or 9/11, media institutions and journalists also act as agents in lifting socially constructed moments (for example, the Copenhagen COP15 of the UN Framework Convention on Climate Change) to varying degrees and ranks of an event through their coverage. A similar dynamic applies to slower processes in nature such as retreating Arctic sea ice that only become events when they are constructed and mediated as such. An example is when press releases conveying satellite data are made public in ways and in a broader social and political context that grabs and further fuels media attention.

In understanding media events, the level and intensity of coverage is an important factor, but qualitative elements also factor in and should be accounted for to grasp the overall picture and ensuing impact in the long term. This echoes McComas and Shanahan's (1999, p. 53) conviction that 'it is not only the frequency of coverage, but also the character and form of that coverage that help to draw public attention'. Over the past two decades, there have been a variety of attempts, such as 'mediatized public crises', 'media disasters' and 'media scandals' to redefine and broaden the concept of media event through the use of related discursive constructs. In an effort to integrate media events and the concept of 'mediatized rituals', Simon Cottle (2008) suggests that the latter term refers to 'those exceptional and performative media phenomena that serve to sustain and/or mobilize collective sentiments and solidarities on the basis of symbolization and a subjunctive orientation to what should or ought to be' (Cottle, 2006, p. 415; see also Couldry, Hepp and Krotz, 2009).

The literature on media events highlights their increasingly global character, both in their production and in their reception (Couldry, Hepp and Krotz, 2009, p. 9; Hepp, 2004). Nationally confined and time-specific eventizations through live broadcasts (such as the Kennedy assassination, Woodstock or royal weddings) are hardly the primary constituents of the definition that should apply to 'media events' of today. More important is the 'mobilization of collective sentiments and solidarities on the basis of symbolization and subjunctive orientation' (Cottle, 2006, p. 415). Such a nuanced approach to media events offers a way to better grasp the 2007 and 2012 sea-ice minima, while a closer analysis of these events provides the empirical grounds to further theorize the mediatization of meta-phenomena such as 'climate change'.[5]

In addition to periodically entering the visible realm of public debate by way of de facto events[6] – quality moments such as headline-friendly disasters and groundbreaking scientific revelations – the climate debate on the whole embodies a truly longitudinal, meta character unlike other events of more momentous nature. As such, and recognizing the need to utilize a multidisciplinary approach to understand climate change in its totality, we propose a conceptual framework within which to approach climate change as a *meta-event* with the Arctic sea-ice minimum constituting a very significant *moment* along the way. As a meta-event, climate change is much more than the sum of its moments. It is akin to other phenomena that are socially and mediatively produced and absorbed as meta-events, such as the fall of the Berlin Wall or, more recently, the Arab Spring (Christensen, 2013). Thus the climate change debate, in and of itself, constitutes an institution that produces a multiplicity of scientific, political, popular and moral positionalities and contingencies across time and space.

Words, images and moments in media reporting

As discussed, the ways in which the media position climate change and its impacts have great significance. In the US context, for instance, skepticism towards a changing global climate is toned down if the label *climate change* is used rather than *global warming*, especially among Republicans (Schuldt, Konrath and Schwarz, 2011). And, just as the language about global warming matters, so do the ways in which global warming and its consequences are displayed through images, pictures, illustrations, maps, animations and so forth. The images surrounding the text are embedded in particular contexts that constitute the basis upon which our understanding and interpretation is built. One such example is discussed by Nina Wormbs in her chapter on the newspaper images illustrating the sea-ice minimum. These illustrations send different messages concerning what is important about the shrinking sea ice in the Arctic. This might be an unintended effect from the point of view of the journalist who used the image to illustrate the text.

Beyond its discursive capacity, and from a factual standpoint, climate change on the whole (and greenhouse effect and global warming, to use its variants) constitutes one of the most significant environmental, social, political and economic phenomena of the twentieth and twenty-first centuries. Unsurprisingly, the past three decades have witnessed an unprecedented level of international scientific and media focus on its causes and consequences. Yet, while the issues at stake are highly complex, reciprocated with a corresponding attention to complexity in the scientific debates as shown by Nilsson and Döscher in this volume, the media as news outlets mainly choose to report moments that are deemed 'worthy' of coverage. The meta-event of climate change and its complexity on the whole lacks visibility and this pattern of media coverage has major consequences for public understanding of scientific and political questions related to the issue. Yet it is also the case that compared to the media discourse of the 1980s and 1990s, the analysis presented by Miyase Christensen in her chapter shows a shift in the mode of address adopted by the media: one that has transformed reporting from a 'they' debate (i.e., climate change as scientific controversy and technocratic problem) to a 'we' debate (that is climate change as a global, *shared* problem of humanity).

In the 2000s (particularly from 2003 onward), international media started to pay attention to the decline in Arctic sea ice as a barometer for global warming, especially through stories with an overall focus on sociocultural, economic and political issues – despite periods of decline in the coverage. In the British press, which plays a leading role in the public debate around climate change, the drop in the Arctic sea ice often hit the headlines to the effect of signifying a climate event with broader consequences. In the case of the Arctic, the media tend to focus on the sea-ice minimum in the month of September, and whether it has broken any records. The coverage of climate change on the whole in the 2000s has followed the common media trend of peaks and valleys with high points during moments of media significance followed by dips in attention. We observe one such peak in 2007 following the release of the IPCC Fourth Assessment Report and of Al Gore's film, *An Inconvenient Truth*, the same year. Later in the fall of 2007, a Nobel Peace Prize was awarded to IPCC and to Al Gore, further highlighting the significance of the climate-change issue in the international media. Yet media's involvement in the climate debate is never unproblematic as the angles and solutions propagated are often entangled with various scientific, political and industry agendas. As we discuss elsewhere (cf. Christensen, 2011), an even higher peak was reached in 2009 due both to the e-mail controversy originating from the Climatic Research Unit and the following COP15 in December 2009.

Mediatization and climate debate

As discussed earlier, as a concept 'media event' cannot be reduced to an event reported in the media for national audiences. It refers to the role of the media, operating in a global and mediatized environment in conjunction with other social actors and how these interactions turn 'something' into a significant moment in time, or turn *anything* into *something*. In media studies, the concept of mediatization is used to describe how social processes are increasingly dependent on mass media and the fact that society is saturated by the media to the point where we cannot think of social institutions and media as separate from one another (cf. Mazzoleni and Schulz, 1999; Hjarvard, 2008). As a baseline, capturing what mediatization entails requires a social analysis of the role of the media and its relations to other actors and institutions. In such an environment, understanding the production and consumption of media representations of climate change (cf. Carvalho, 2005) becomes crucial. Politicians and other social stakeholders pay particular attention to the ways in which the media treat issues such as climate change, and, the media constitute the primary sources on which the public depends for acquiring information and forming opinion (even though it is certainly not the case that the encoded/intended message and the decoded/received meaning are the same).

It is not only politics that is mediatized (Meyer, 2002; Mazzoleni and Schulz, 1999; Schulz, 2004). On the whole, the media play an increasingly significant and complex role in understanding a whole range of social phenomena, from everyday micro and private issues such as family and friend networks to global political and economic processes. New communications technologies and globalization have made the media and how different actors use and relate to the media an entrenched characteristic of contemporary culture and society. As suggested by Livingstone,

It seems that we have moved from a social analysis in which the mass media comprise one among many influential but independent institutions whose relations with the media can be usefully analysed, to a social analysis in which everything is mediated, the consequence being that all influential institutions in society have themselves been transformed, reconstituted, by contemporary processes of mediation. (2009, p. 3)

Influential institutions here refer to the key constituents of social life that guide our norms and expectations of one another. Recently there have been more in-depth analyses of various aspects of mediatization, hence new formulations ensued. Krotz (2008), for instance, takes mediatization to signify a meta-process, akin to globalization. It is an intrinsic aspect of modernity (Krotz, 2008, p. 23). As such, the media play a crucial role not only by way of providing a 'communicative space' for the re/presentation of issues of scientific and political importance, but by way of bringing a 'popular' dimension to the overall realm of knowledge production, policymaking and public debate. In news research, mediatization is used in order to explain how public understanding of issues of political, scientific and social significance is shaped.

We should also note that as regards mediatization and the role of the media in the climate-change debate, a dialectical understanding of the media as having significant *social influence* while at the same time being *socially shaped* remains essential. Media norms of impartiality – providing coverage of pros and cons – often distorts the representative picture of scientific conflict or consensus. Policy statements in the area of climate change increasingly build on scientific consensus, such as the statements from the IPCC, even if the realm of public policymaking itself often lags behind scientific advice about the need for action. The media, however, rarely rely on the strength of scientific evidence in assigning appropriate space for different voices, especially not where there is a scientific conflict. Instead, the media logic tends to provide favorable treatment to statements that go against consensus; such views are more likely to make the headlines. The public can therefore not safely rely on the media to get a fair picture of scientific consensus on politically contentious issues (cf. Oreskes and Conway, 2010).

Topical Multiplicity

Because mediatization is a prominent feature of late-modern societies, media accounts provide a useful source of information, a mirror reflecting how our understanding of the world we live in is shaped. More specifically, they provide a starting point for pinning down and tracing which worldviews are foregrounded, and why, and what other ways of seeing have been excluded or deemed less prominent. The in-depth analysis of the media accounts relevant to the sea-ice minimum presented by Christensen in this volume highlight two such trends. One is that the sea ice becomes a representational venue for a multiplicity of topics that go far beyond issues of climate change in the Arctic: *topical multiplicity*. The other trend has to do with how social and material processes of different scales – from local to global and vice versa – converge in the same story frames: *scalar transcendence*.

Topical multiplicity helps to explain how shrinkage of the Arctic sea ice has been covered in a broader sense and how media reports on this issue provided a scope that surpassed the behavior of the ice, the climate, or nature in general. As suggested at the beginning of this chapter, new commercial possibilities in relation to shipping are one such example. Energy and mineral resources also figure in the discussions about the Arctic and its future. In this context, it is worthwhile to highlight that our point of departure is that the environment is a 'human product' with a history (Sörlin and Warde, 2009, p. 3). As such it can and should be analyzed and studied with the methods and theories devised for analyzing human aspects, that is, social sciences and humanities, including social studies of science and technology. Thus the conceptual frames of *topical multiplicity* and the linking of issues across scales in the media – *scalar transcendence* – provide standpoints for analyzing how we construct the environment today.

Topical multiplicity is not a concept that only has relevance in terms of media stories and news coverage. It is also relevant for understanding the context of environmental policy. Environmental policy deals with much broader issues than questions directly related to nature. The concept of sustainable development is a case in point with its three at times conflicting pillars of environmental, social and economic development (Owens, 2003). The focus of global environmental governance has also expanded significantly since the first UN Conference on the Human Environment in 1972 and now includes broad concerns about poverty reduction, equity and allocation of resources (Biermann et al., 2009). The broadened agenda has raised the stakes in negotiations and made environmental policy increasingly political, which has also made it difficult to arrive at any far-reaching agreements, as evidenced by the climate negotiations and also the Rio+20 process. Recently the issue of global resource scarcity has re-entered the environmental policy discussions with implications for the political climate. According to Andrews

Speed et al. (2012), global resource demand is likely to accelerate in the next 10 to 20 years, at the same time as resources such as water, energy, land and food and minerals are increasingly connected, creating much more complex interdependencies between issue areas than when resource scarcity was discussed in the 1970s. In this context, the environment – globally and in the Arctic – is quickly moving from being an area of low politics suitable for compromises to an area of high politics where national self-interests are foregrounded (Nilsson, 2012b). How issues are framed plays an important role in environmental politics: whether the dominant framing invites conflicting interests or suggestions for mutually beneficial cooperation (Underdal, 2001). Topical multiplicity, then, is evident in multiple concurrent framings and points to the role of the media in highlighting certain ways of viewing the world.

Scalar Transcendence

The issue of *scalar transcendence* links to how we frame changes in the Arctic in relation to scale, and highlights not only how sea-ice decline is reported but also how it is observed. For centuries peoples of the region have been observing the sea ice in the Arctic. In the wake of remote sensing satellites in the late 1970s, scientists became able to generate vast amounts of data of the circumpolar sea-ice extent – a commonly –referred to indicator of climate change in the polar region. Other important indicators are the thickness and age of the ice. The development of satellite technology has brought with it a change in scale from the local to the regional and ultimately to the global that had a predecessor in the collection of data sets for meteorology (Edwards, 2010). Globalization of science dates back to an earlier era, where the international polar years of 1882–1883 and 1932–1933 and the International Geophysical Year 1957–1958 provided milestones for understanding various global processes and placing the polar regions into a global context, while at the same time serving as spaces for nationalism and diplomacy (Launius et al., 2010).

Science and Technology as a Social Process

Just as mediatization is a dual process – both affecting and being affected by social processes – so is the dialectic between the science and technology of climate change and the public debate on its origins, its features and its consequences. The public debate on climate change and the science and technology of climate change are co-produced, to use Sheila Jasanoff's term (Jasanoff, 2004). Jasanoff belongs to a tradition of scholarship emanating from sociology and history investigating the epistemological and social status and development of science and technology in society. Sociologists and historians of science, as well as scientists themselves (Kuhn, 1962), started to question the absolute-truth claims imbued into scientific findings, showing that in fact how science is made and valued is a social and cultural matter, contextually situated, historically dependent (Latour and Woolgar, 1979; Shapin and Shaffer, 1985; Haraway, 1989). In parallel, research on the interplay between how technology is both affected by and is affecting society was influenced by this constructivist ideal, resulting in a more nuanced picture of the technological society and its development (Hughes, 1983; Bijker, Hughes and Pinch, 1987). Moreover, the line between what was to be termed science and what one could call technology was blurred and questioned (Latour, 1987).

The spectrum of possible positions on science and its construction was enlarged and the relativist turn at times overshadowed the fact that most researchers in the field of science and technology studies (STS henceforth) actually never questioned the existence of a reality and of observable facts, even though the term itself is contested. However, even in mainstream STS, the art of the disenchantment of science and technology was so well performed that public discourse as well picked up and absorbed the central tenant of constructivism, challenging the truth claims made by the scientific community. Paul Edwards has argued that perhaps the success of STS in this broader context has done the climate issue a disservice since it is possible to question and overthrow scientific findings on precisely the ground that science is a construct and as such it should be questioned (Edwards, 2010).

Indeed, to question a result is fundamental to science and therefore doubt should be high. This means that the scientific community is constantly trying to corroborate and interrogate the existing view of things. However, it is only when this process actually leads to a *possible* re-evaluation of an established fact that it is of any interest for media to report on it. Subsequently, studies confirming an established position will not get media attention, even though they might be much larger in numbers, while those challenging the established view will get a disproportionally higher level of publicity. This logic can also be utilized for the sole purpose of undermining the public view on a particular scientific consensus, thereby obstructing science-based policy (Oreskes and Conway, 2010).

How we look upon science is hence important for how we understand and evaluate scientific findings on the state of the environment. Moreover, there is hardly any science that does not also involve technology. As hinted above, there are even those advocating the collapse of the terms into techno-science, signaling the intertwined nature and mutual dependence of the knowledge areas (see, for example, Latour, 1987). It is by technological means that we know things about our changing climate, and this has been true for a long time. Field stations have proven essential to diachronic monitoring, statistics have been central to distinguish trends, surveillance equipment like remote sensing satellites have enabled the collection of data, and computers have made it possible to collate and make sense of vast amounts of data. Climate science as we know it would be impossible without this technological infrastructure (Edwards, 2010), even though recent studies have also proven the importance of traditional knowledge (Krupnik et al., 2010).

However, our understanding of the role of technology in the production of knowledge in climate science is not a given. The understanding has shifted from technology being a mere tool of science, transparent and with no agency of its own, to technology determining any scientific claim. As always, the most fruitful position might be one acknowledging both the enabling and limiting factors of technology, understood as the latter affording certain outcomes but not others. If we take satellites, for example, they provide us with almost unique data that would have been impossible to attain with other means. It is comprehensive and reasonably comparable over time. However, at the same time, these data sets are also limited in some respects since their resolution in time and space is given by the orbits of the satellites, the functionality of the information-gathering equipment, and the possibility to download data to ground stations on Earth. In this volume we also take interest in how these data sets are transformed and visualized into images that are easier to grasp than the numbers themselves (see Wormbs in this volume).

Remote sensing images come in different forms and can be local, regional and global at the same time depending on the resolution of the data. When surveying sea ice in the Arctic, a regional or global perspective has been useful since it is the change over an entire region over time that has been of central interest. Some of the images that are produced in order to visualize the change resemble well-known, iconic images of the Earth and make us connect the ongoing change to things we already know about the planet, a matter discussed in the chapter by Wormbs. The process by which these images become iconic is, of course, one where the media are instrumental (Poole, 2008). The remote sensing images of the shrinking sea ice in the Arctic are then not just products of science but are enabled by technology and given meaning through complex cultural and societal processes over time.

Such cultural and social processes also play a role in defining the projections we use to view the world. Take, for example, the standard mapped image of Earth. When polar regions are not in focus, such maps provide a distorted picture of distance and areas. When the Arctic is an ice-covered area of little global interest, this may not matter, but when attention and interests shift, we also see more and more images with a polar projection where patterns of trade and other economic and political activities can be represented in ways that foreground certain interests. They can be placed in contrast to local descriptions of the ice, which have more immediate impact on different activities, whether they are commercial enterprises relating to resource extraction and shipping, or local livelihood, as highlighted in Chapter 6 by Henry Huntington.

Arctic region-building

The close links between science, technology and social structures also include the relationship to patterns of political cooperation. Climate science is a particularly visible case in point where the growth of the international science of climate change has been closely interrelated with the growth of a global political society and the United Nations system after World War II (Miller and Edwards, 2001). In a similar vein, we argue that the current understanding of Arctic sea ice needs to be analyzed in the context of Arctic political cooperation as it has emerged in the last 25 years.

Less than two decades ago, few people other than its residents, the military and polar scientists took notice of the Arctic. It was a region of political and military tensions between NATO and the Warsaw Pact (Heininen, 2004) and mostly left out of the media limelight. Today the Arctic is often spelled with a capital A, indicating that it has gained region status and an identity of its own. The first turning point came in 1987 with the so-called Murmansk speech during which the Soviet president Michael Gorbachev called for making the Arctic a zone of peace and drew attention to the Arctic environment and its indigenous people (Åtland, 2008; Scrivener, 1989). The speech set in motion a process that has been described as region-building (Keskitalo, 2004) and which included the creation in 1991 of a high-level forum for international cooperation among eight Arctic states as codified in the Arctic Environmental Protection Strategy (Young, 1998). In 1996 the Arctic Council was created in furtherance of the broader policy goal of sustainable development. The working groups of the Arctic Environmental Protection Strategy and later the Arctic Council have brought the scientific and policy spheres together in a way that has made the Arctic Council a cognitive forerunner in defining the Arctic as a region in the context of international policy of chemicals and as regards knowledge about the impact of climate change (Nilsson, 2012a).

While the more open political climate in the Arctic created new opportunities for international scientific cooperation, scientific research about the Arctic Ocean as such has a much longer history. Whaling exploration and efforts to find new sea routes started in the seventeenth century. Since the mid-nineteenth century, nationalistically inspired explorations brought images of the northern seascape to the attention of people outside the region. Some of these were also instrumental for our current understanding of the Arctic. To mention just one example relevant for this book, the famous drift of Fridjof Nansen's vessel *Fram* from 1893 to 1896 revealed then-unknown large-scale movements of the ice and water masses in the Arctic Ocean (Nansen, 1897). Later, in the mid-twentieth century, the 1957/1958 IGY included drifting polar-ice stations to collect data on atmosphere, ocean and ice conditions, an idea initially proposed by Nansen and first implemented by the Soviet Union in 1937 (Barry, 1983). Although civilian research about the Arctic Ocean continued during the 1900s, it retained its exploratory character until 1980, at least in the West (Rudels et al., 2012). For the military, a better understanding of Arctic conditions led to investments in research much earlier as knowledge on the condition of the Arctic Ocean was crucial to many of the military operations during both the World War II and the Cold War. In the West, in the United States in particular, this led to a revival of earth sciences, including oceanography and attention to the polar region (Doel, 2003).

More recent developments, starting in the 1980s, can be linked to increasing interest in understanding the Arctic as part of a global climate system, increasing international coordination of science, and to new technologies for gaining access to the Arctic. In 1980, a new era of civilian research about the Arctic Ocean started with the introduction of icebreakers as research platforms. The 1980s also brought the first large multinational research program on Arctic sea ice: the Marginal Ice Zone Experiment in 1983 and 1984 (Rudels et al., 2012, p. 124). However, the political context, including the extended national sovereignty of the continental shelf, limited access to scientifically relevant parts of the Arctic (Theutenberg, 1982).

The shift in the political climate in the Arctic in the late 1980s created a setting with new possibilities for what became a wave of internationally coordinated endeavors that have played a major role in advancing the knowledge about the circumpolar north in general and the Arctic Ocean in particular. This included the Arctic Climate System Study which set out to document the state of the Arctic pack ice and study the feedback between the sea ice and other elements of the climate systems, motivated by the need for a better empirical foundation in the work on improving the computer models that are used for simulating global climate change (Melling, 2012, p. 29). It was followed by other coordinated research efforts to understand ice and snow in all its forms and their role in the Earth system (Steffen et al., 2012, p. 437) that were closely linked to the World Climate Research Programme. This, in turn, can be seen as an offspring of the growth of global international society after World War II in its relation to scientific cooperation. European political integration has also played a role in advancing knowledge about Arctic sea ice, for example by the EU project DAMOCLES, which was specifically concerned with the potential for a significantly reduced sea-ice cover.[7] These and other coordinated research efforts have in the past 30 years built an extensive international infrastructure for civilian research about the Arctic sea ice. Without them, and the political climate and cooperative ventures that made them possible, we may not have noticed the sea-ice minimum of 2007.

The new wave of coordination of polar research that took off after the end of the Cold War culminated with the International Polar Year 2007–2008. Its first year coincided with the Arctic sea ice reaching its unprecedented minimum in 2007. Akin to previous international polar years, the purpose of the IPY 2007–2008 was to create enhanced observational systems, new research facilities and infrastructure that could be shared internationally (Krupnik et al., 2011, p. xix). As in the case of the preceding polar years, including the International Geophysical Year 1957–1958, the driving forces behind the planning and execution of IPY 2007–2008 are best understood in the context of the relationship between science and politics. One notable aspect in the case of IPY 2007–2008 was that it coincided with a growing awareness about global climate change and its impact as a 'long-term crisis for all humanity' (Bravo, 2010, p. 447). That the Arctic sea ice was diminishing was starting to become apparent, and understanding these changes was one of the key priorities already in the early planning of the IPY. In the end, the IPY featured more than 30 projects focused on the polar sea ice (Krupnik et al., 2010, p. 3).

The political context that made the most recent polar year different from its predecessors and most probably influenced its role as part of a media event was that environmental change was by now well entrenched as a global and international concern, rather than merely a national or local issue, and environmental governance increasingly an area of global politics (Nilsson, 2012b). Another political difference to preceding polar years was the increasing recognition of the rights of indigenous peoples and a higher level of attention paid to their traditional knowledge. Although originally conceived mainly as an initiative for the geophysical sciences, the IPY 2007–2008 came to include attention to cultural, historical and social processes that shape circumpolar human society (Krupnik et al., 2010, p. 5). Studies of the sea ice came to play a substantial role in these efforts, for example in the project SIKU (Sea Ice Knowledge and Use). *Siku* is also the Inuit word for sea ice (Krupnik et al., 2010, p. 5).

Last but not least, a cooperation-based political climate made agreements on sharing of observations and other scientific data much more feasible than this would have been at the height of the Cold War. Data sharing has been one of the purposes of international polar years since the first one in 1882–1883, but the combination of new technologies, a geopolitical climate that favored cooperation in the Arctic, and recognition of knowledge traditions other than the geophysical sciences has set the stage for a different level of data sharing, which we argue also plays a role in relation to media images of Arctic climate change.

Natural science pictures of Arctic sea ice

When plans for the IPY were made, the Arctic Climate Impact Assessment had already drawn attention to the potential risk of diminishing sea ice in the Arctic (ACIA, 2005). However, compared to what actually happened in the following years, those estimates proved to be conservative. The need for a better understanding of the impacts of global climate change on ice and snow in the Arctic was recognized by the Arctic Council and led to a new assessment, which was officially approved in the spring of 2007 with results presented in the spring of 2011 Snow, Water, Ice and Permafrost in the Arctic (SWIPA). The report contains a chapter specifically focusing on Arctic sea ice (Meier et al., 2011), on which we have based the following very brief synthesis of scientific understanding of Arctic sea-ice dynamics.

The SWIPA authors writing about the sea ice conclude that the decline in Arctic summer sea-ice extent has accelerated over the past 10 years and that ice-free summer conditions are likely over most of the Arctic basin by mid-century or earlier. They also report that the ice is now thinner and younger with the decline of some of the old, multi-year ice.

In fact, observations indicate that most of the oldest ice types (more than five years old) have been lost. The authors conclude that it is likely that the Arctic will become increasingly taken over by first-year ice, which represents a different type of physical environment for animals as well as ships and industrial activities (Meier et al., 2011). It is, for example, thinner, saltier and more mobile than the aged pack ice that has been shaped and often reshaped over several years' time.

The fundamental driving force yielding ice loss is the increase in temperature. While easy to understand at one level, the links between a warmer climate and sea-ice loss are quite complex, as are the mechanisms that link temperatures in the air with extent and thickness of the ice. This means that changes in ice cover are not likely to directly follow a curve of changing temperatures. Recent understanding instead points to feedback loops that speed up the melting at certain times in the year. For example, in the spring, warmer weather has led to earlier melting of the snow, which otherwise would reflect away the energy from the sun. With the snow gone, the melting accelerates, and the exposed dark water surfaces accelerate it even further. One of the key findings of the SWIPA report was that such feedback loops may be speeding up the ice loss overall beyond what would have been expected had one looked at temperature effects alone. Moreover, the Arctic sea ice is in constant movement. When the ice gets thinner it moves even more, which exposes yet more areas of dark open water that can effectively absorb the solar energy.

In addition to higher temperatures and more energy absorbed from the sun, changes in wind patterns have also played an important role for the changes in Arctic sea ice. This appears to be the case especially for the decline of the multi-year ice, which has moved out of the Arctic unusually fast. The latest science does not fully explain why the Arctic sea ice declined much faster in 2007 than had been expected, but the understanding that has emerged in recent years hints at the complexity involved (AMAP, 2011, p. 9–16).

Observations of the sea ice by indigenous people also point to a much more complex picture than is usually visible in media accounts, as elaborated by Huntington in his chapter. Eicken (2011) also highlights the importance of how the local scale of indigenous observation can reveal features that are invisible from satellite data but nevertheless crucial both for the users of the ice (or the open water between) and for those trying to understand how and why the sea ice is changing. Indigenous knowledge also provides new insights related to the temporal scale of change, both by extending the record of observation further back in time and by providing observations of rare events, and events that may appear as invisible or inconspicuous when using scientific methods of observation.

These more complex stories about the ice have not (yet) become prominent in media coverage about Arctic change. Understanding what was missing from the media stories related to the sea-ice shrinkage is as important as accounting for what was included in the news coverage of the same phenomenon.

Chapter structure and concluding remarks

The current volume brings together different scholarly outlooks rooted in disciplines such as social theory, media studies, environmental history, oceanography and climate modeling, science and technology studies, political science and anthropology. Apart from this Introduction and the concluding chapter, the volume contains six chapters that offer different 'ways of seeing' Arctic climate change. Miyase Christensen's chapter is based on a qualitative frame analysis of news reporting on the sea-ice minimum over a number of years in three high-ranking and influential daily newspapers: the Swedish *Dagens Nyheter*, *The Guardian* and *The New York Times*. Christensen finds that the quality press reporting on the sea-ice minimum did not stress scientific conflict or controversy but rather discussed the event as a scientific truth showing that the globe is indeed warming. Stories containing multiple dimensions and topics were told in parallel to that of the minimum itself, incorporating different scales, reaching from the local to the global. The following chapter by Nina Wormbs also has news reporting in focus, but more specifically addresses how different images are used to illustrate the written text. As the scientific reporting relied heavily on remote sensing data from the late 1970s, the transformation of those data into images and their subsequent use in reporting is the main interest of this chapter. As Wormbs points out, the choice of illustration carries more weight than might be anticipated at first sight.

The third chapter by Sverker Sörlin and Julia Lajus approaches the sea ice as a historic phenomenon. Measurements of the ice have been carried out before the advent of satellites and this chapter traces the politics of sea-ice science, emphasizing the significance of narrative construction as well as scientific endeavors. Interestingly enough, natural scientists are now also conducting historical research in order to get longer time series and deepen our understanding of the sea ice over time (Walsh et al. 2011).

The following chapter connects to those long-term trends by analyzing how scientists have discussed variability and change when it comes to Arctic climate change. Annika E. Nilsson and Ralf Döscher conclude that scientific writers still focus on the *variability* of climate, not as a contrast or in opposition to the popular discourse and policy consensus that climate change is real. Rather, their analysis reveals two parallel discourses in the scientific literature only one of which has reached a broader audience. The scientific discussion on natural variability that has been neglected by the media may be vitally important for issues related to adaptation to climate change. But in mediatized frames focusing on climate controversy, this has also been difficult to handle and carried the risk of diluting the overall message that the climate is actually changing.

The penultimate chapter deals with the local perspective and how the minimum was experienced by the communities along the coast of northern Alaska. Henry Huntington finds that the local communities still fished and hunted and that the ice did come back when winter returned. The story of those living where the ice melted did not make the headlines. Such local perspectives not only bring out the large differences between various locations, but also the role of ice for local people and that the ice still returns in the winter. Finally, Dag Avango and Per Högselius trace the efforts to extract natural resources, such as fossil fuels, from the Arctic before the region started to melt. Resource exploitation has been going on for a long time in this region even though the attention has shifted from coal to oil and gas. In their chapter, Avango and Högselius show how ice can sometimes be called upon as an asset, rather than an obstacle, to extraction, illustrating the role of technology not just as a practical, but a rhetorical, tool. They argue that resource exploitation would have continued even without the retreating sea ice and through their examples they forcefully undress the climate-deterministic rhetoric behind the current discourse.

In sum, the current volume is an attempt to analyze Arctic climate change and its local/global dimensions from an encompassing, multidisciplinary and complex perspective. It highlights how mediated stories connected to changing political structures transform the Arctic sea ice from being a local concern for Arctic residents to a global concern with links to both international climate politics and to larger geopolitical changes. The analyses we pursue resonate with scholarly trends of the past three decades in the social sciences and humanities towards identifying worldwide social phenomena and change in the paradigmatic scope of globalization. The latest IPY was different from the earlier ones in that social sciences and humanities were incorporated to a much larger extent than ever before (Shadian and Tennberg, 2009, p. 2). Historical and anthropological projects focusing on sea ice (Krupnik et al., 2010) contributed to the knowledge base otherwise built and appropriated by natural scientists. This broadening of the investigation of the cryosphere also reflects a wider interest in climate change from the view of the social sciences and humanities. In the Anthropocene – as a new geological era where human society is a significant driving force – it becomes crucial to understand the social forces that not only shape nature but also how those social forces shape and are shaped by our understanding of the environment.

In September 2012, the National Snow and Ice Data Center released information about the Arctic sea ice reaching below the 2007 extent. Among the stories circulated through international news outlets are those of shipping developers eyeing routes through the melting Arctic, and China's recent use of the Northern Sea Route to reach Iceland with its vessel *Snow Dragon*: 'Arctic Ocean here we come!'[8]

Trade remains vital to global capitalism and certainly to growing economies such as that of China. Media reporting about the Arctic today, that goes beyond scientific controversy and scalar singularity, is testimony to the fact that in the Anthropocene we are faced globally with a new Arctic in both material and discursive terms. Grasping this "new" old region and the vast changes the Arctic sea ice is presenting pose disciplinary *and* post-disciplinary challenges that lie ahead of us.

Notes

1. http://www.guardian.co.uk/environment/2011/sep/11/arctic-ice-melting-at-fastest-pace.
2. http://www.americanpolar.org/2011/08/19/speed-record-on-northern-sea-route/.
3. http://www.nytimes.com/2011/10/18/business/global/warming-revives-old-dream-of-sea-route-in-russian-arctic.html?_r=1&pagewanted=1&hp.
4. http://nsidc.org/news/press/2007_seaiceminimum/20071001_pressrelease.html.
5. This is not to suggest that the sea-ice minima necessarily led to solidarity. But they did trigger an understanding of urgency and global destiny with the imagery of the Arctic solidifying the idea of the polar region being the bellwether of change.
6. Taking event in its simplest, most mundane sense.
7. http://www.damocles-eu.org/.
8. http://iphone.france24.com/en/20120826-arctic-melts-developers-new-shipping-northern-sea-route-russia-china-ice-loss.

References

ACIA (2005) *Arctic Climate Impact Assessment* (Cambridge: Cambridge University Press).

AMAP (2011) *Snow, Water, Ice and Permafrost in the Arctic (SWIPA) 2011: Climate Change and the Cryosphere* (Oslo: Arctic Monitoring and Assessment Programme).

Andrews-Speed, P., R. Bleischewitz, T. Borsma, C. Johnson, G. Kemp and S. D. VanDeveer (2012) "The Global Resource Nexus. The Struggles for Land, Energy, Food, Water, and Minerals". Washington DC: Transatlantic Academy.

Åtland, K. (2008) 'Mikhail Gorbachev, the Murmansk Initiative, and the Desecuritization of the Interstate Relations in the Arctic', *Cooperation and Conflict: Journal of the Nordic International Studies Association*, 43:3, 289–311.

Barry, R. G. (1983) 'Arctic Ocean Ice and Climate: Perspectives on a Century of Polar Research', *Annals of the Association of American Geographers*, 73:4, 485–501.

Biermann, F., et al. (2009) *Earth System Governance. People, Places and the Planet. Science and Implementation Plan for the Earth System Governance Project*, IHDP, Earth System Governance Project, Bonn. Earth System Governance Project Report no 1. IHDP Report no 20.

Bijker, W. E., T. P. Hughes and T. J. Pinch (eds) (1987) *The Social Construction of Technological Systems: New Directions in the Sociology and History of Technology* (Cambridge, MA: MIT Press).

Bravo, M. T. (2010) 'Epilogue: The Humanism of Sea Ice', in I. Krupnik, C. Aporta, S. Gearheard, J. L. Laidler, and L. Kielsen Holm (eds), *SIKU: Knowing Our Ice Documenting Inuit Sea Ice Knowledge and Use* (Dordrecht: Springer, 445–52).

Carvalho, A. (2005) 'Representing the Politics of the Greenhouse Effect', *Critical Discourse Studies*, 2:1, 1–29.

Christensen, M. (2011) 'Discursively Shaping the Environment: Swedish National and Regional Media Coverage of the United Nations Climate Change Conference in Copenhagen ("COP15")', Paper presented to the Global Communication and Social Change Division of the International Communication Association (ICA) Conference, May 2011, Boston.

Christensen, M. (2013) 'New Media Geographies and the Middle East', *Television and New Media* 14:267.

Cottle, S. (2006) *Mediatized Conflict: Developments in Media and Conflict Studies* (http://www.findanexpert.unimelb.edu.au/individual/publication57223) (Open University Press).

—— (2008) 'Mediatized Rituals: A Reply to Couldry and Rothenbuhler', *Media, Culture & Society*, 30:1, 135–40.

Couldry, N. (2003) *Media Rituals: A Critical Approach* (London: Routledge).

Couldry, N., A. Hepp and F. Krotz (eds) (2009) *Media Events in a Global Age* (London: Routledge).

Dayan, D., and E. Katz (1992) *Media Events: The Live Broadcasting of History* (Cambridge, MA: Harvard University Press). Reprinted by Harvard University Press (1994).

Doel, R. E. (2003) 'Constituting the Postwar Earth Sciences: The Military's Influence on the Environmental Sciences in the USA after 1945', *Social Studies of Science*, 33, 635.

Edwards, P. N. (2010) *A Vast Machine: Computer Models, Climate Data, and the Politics of Global Warming* (Cambridge, MA: MIT Press).
Eicken, H. (2010) 'Indigenous Knowledge and Sea Ice Science', in I. Krupnik et al. (eds), *SIKU: Knowing Our Ice. Documenting Inuit Sea-Ice Knowledge and Use* (Dordrecht: Springer, 2010, 357–76).
Fiske, J. (1994) *Media Matters: Everyday Culture and Political Change* (Minneapolis: University of Minnesota Press).
Haraway, D. J. (1989) *Primate Visions: Gender, Race, and Nature in the World of Modern Science* (New York: Routledge).
Heininen, L. (2004) 'Circumpolar International Relations and Geopolitics', in AHDR (ed.), *Arctic Human Development Report* (Akureyri, Iceland: Stefansson Arctic Institute).
Hepp, A. (2004) *Netzwerke der Medien. Medienkulturen und Globalisierung* (Wiesbaden: VS).
Hjarvard, S. (2008) 'The Mediatization of Religion: A Theory of the Media as Agents of Religious Change', in *Northern Lights: Yearbook of Film & Media Studies* (Bristol: Intellect Press).
Huebert, R., H. Exner-Pirot, A. Lajeunesse and J. Culledge (2012) *Climate Change & International Security: The Arctic as a Bellwether* (Arlington, VA: Center for Climate and Energy Solutions). (http://www.c2es.org/publications/climate-change-international-arctic-security, date accessed 20 February 2012.)
Hughes, T. P. (1983) *Networks of Power: Electrification in Western Society, 1880–1930* (Baltimore: Johns Hopkins University Press).
IPCC (2007). 'Climate Change 2007: The Physical Science Basis. Summary for Policymakers. Contribution of Working Group I to the Fourth Assessment Report of the Intergovernmental Panel on Climate Change'. Geneva: Intergovernmental Panel on Climate Change. www.ipcc.ch.
Jasanoff, S. (ed.) (2004) *States of Knowledge: The Co-Production of Science and the Social Order* (London: Routledge).
Katz, E. (1980) 'Media Events: The Sense of Occasion', *Studies in Visual Anthropology*, 6, 84–9.
Keskitalo, C. (2004) *Negotiating the Arctic: The Construction of an International Region* (London: Routledge).
Krotz, F. (2007) 'The Meta-Process of "Mediatization" as a Conceptual Frame', *Global Media and Communication*, 3:3, 256–60.
—— (2008) 'Media Connectivity: Concepts, Conditions, and Consequences', in A. Hepp, F. Krotz and S. Moores (eds), *Network, Connectivity and Flow: Key Concepts for Media and Cultural Studies* (New York: Hampton Press).
Krupnik, I., et al. (eds) (2011) *Understanding Earth's Polar Challenges: International Polar Year 2007–2008. Summary by the IPY Joint Committee*. University of the Arctic publications series (4). University of the Arctic and ICSU/WMO Joint Committee for International Polar Year 2007–2008, Rovaniemi, Finland.
Kuhn, T. S. (1962) *The Structure of Scientific Revolutions* (Chicago: University of Chicago Press).
Latour, B. (1987) *Science in Action: How to Follow Scientists and Engineers Through Society* (Cambridge, MA: Harvard University Press).
Latour, B and S. Woolgar (1979) *Laboratory Life: The Social Construction of Scientific Facts* (Beverly Hills: Sage).

Launius, R., J. Fleming, and D. DeVorkin (eds) (2010) *Globalizing Polar Science* (New York: Palgrave Macmillan).

Livingstone, S. (2009) 'On the Mediation of Everything' [http://eprints.lse.ac.uk/21420/] ICA Presidential address 2008, *Journal of Communication*, 59:1, 1–18.

Mazzoleni, G., and W. Schulz (1999) '"Mediatization" of Politics: A Challenge for Democracy?', *Political Communication*, 16:3, 247–61.

McComas, K., and J. Shanahan (1999) 'Telling Stories about Global Climate Change: Measuring the Impact of Narratives on Issue Cycles', *Communication Research*, 26:1, 30–57.

Meier, W. N., S. Gerland, M.A. Granskog and J. R. Key (2011) 'Sea Ice', in AMAP, *Snow, Water, Ice, and Permafrost in the Arctic (SWIPA) 2011: Climate Change and the Cryosphere* (Oslo: Arctic Monitoring and Assessment Programme).

Melling, H. (2012) 'Sea-Ice Observation: Advances and Challenges', in P. Lemke and H.- W. Jacobi (eds), *Arctic Climate Change The ACSYS Decade and Beyond* (Dordrecht: Springer, 27–115).

Meyer, T. (with L. Hinchman) (2002) *Media Democracy: How the Media Colonise Politics* (Cambridge: Polity).

Miller, C. A., and P. N. Edwards (eds) (2001) *Changing the Atmosphere: Expert Knowledge and Environmental Governance* (Cambridge, MA: MIT Press).

Morley, D.S. (2009) 'For a Materialist Non-Media-Centric Media Studies', *Television and New Media*, 10:1, 114–16.

Nansen, F. (1897) *Fram over Polhavet: den norske polarfærd 1893–1896* (Kristiania: Aschehoug).

Nilsson, A. E. (2012a) 'Knowing the Arctic: The Arctic Council as a Cognitive Forerunner', in T. Axworthy, T. Koivorova and W. Hasanat (eds), *The Arctic Council: Its Place in the Future of Arctic Governance* (Toronto: Munk-Gordon Arctic Security Program) (http://gordonfoundation.ca/publication/530, date accessed 15 June 2012).

—— (2012b) 'The Arctic Environment – from Low to High Politics', *Arctic Yearbook 2012*. Northern Research Forum and University of the Arctic. On-line http://www.arcticyearbook.com/.

Oreskes, N. and E. M. Conway (2010) *Merchants of Doubt: How a Handful of Scientists Obscured the Truth on Issues from Tobacco Smoke to Global Warming* (London: Bloomsbury Press).

Owens, S. (2003) 'Is There a Meaningful Definition of Sustainability?', *Plant Genetic Resources*, 1, 5–9.

Poole, R. (2008) *Earthrise: How Man First Saw the Earth* (New Haven, CT: Yale University Press).

Rudels, B., L., et al. (2012) 'Observations in the Ocean', in P. Lemke and H. W. Jacobi (eds), *Arctic Climate Change: The ACSYS Decade and Beyond* (Dordrecht: Springer, 117–98).

Schuldt, J. P., S. H. Konrath and N. Schwarz (2011) '"Global Warming" or "Climate Change"? Whether the Planet Is Warming Depends on Question Wording', *Public Opin Q*, 75:1, 115–24.

Schulz, W. (2004) 'Reconstructing Mediatization as an Analytical Concept', *European Journal of Communication*, 19:1, 87–101.

Scrivener, D. (1989) *Gorbachev's Murmansk Speech: The Soviet Initiative and Western Response* (Oslo: Norwegian Atlantic Committee).

Shadian, J. and M. Tennberg (eds.) (2009) *Legacies and Change in Polar Sciences. Historical, Legal and Political Reflections on the International Polar Year* (Farnham: Ashgate).

Shapin, S., and S. Schaffer (1985) *Leviathan and the Air-Pump: Hobbes, Boyle, and the Experimental Life: Including a Translation of Thomas Hobbes, Dialogus physicus de natura aeris by Simon Schaffer* (Princeton: Princeton University Press).

Steffen, K. et al. (2012) 'ACSYS: A Scientific Foundation for the Climate and Cryosphere (CliC) Project' in P. Lemke and H.-W. Jacobi (eds), *Arctic Climate Change. The ACSYS Decade and Beyond* (Dordrecht: Springer), 437–459.

Sverker, S. and P. Warde (eds) (2009) *Nature's End: History and the Environment* (London: Palgrave Macmillan)

Theutenberg, B. J. (1982) 'Polarområdena – Politik och folkrätt', in Swedish Royal Academy of Sciences (ed.), *Polarforskning. Förr, nu och i framtiden* (Stockholm: Swedish Royal Academy of Sciences), 40–55.

Underdal, A. (2001) 'One Question, Two Answers', in E. L. Miles (ed.), *Environmental Regime Effectiveness: Confronting Theory with Evidence* (Boston: MIT Press), 3–46.

Walsh, J., F. Fetterer, W. Chapman and A. Tivy, 'Back to 1870: Plans for a Gridded Sea Ice Product Based on Observations', presented at the 12th meeting of the International Ice Charting Working Group, October 17–21, 2011, Cambridge, U.K.

Young, O. R. (1998) *Creating Regimes: Arctic Accords and International Governance* (Ithaca, NY: Cornell University Press).

2
Arctic Climate Change and the Media: The News Story That *Was*

Miyase Christensen

Introduction

In a mediatized social environment, national and international news outlets and other popular information sources are central actors. They influence not only public debate, but also how politicians, representatives of the business community and other power-brokers position themselves. The increased political and media profile of climate change over the past decade has helped revise earlier discourses and abstract imaginaries of global warming, greenhouse effect and ozone depletion into more concrete social concerns associated with the changing Arctic and planetary future.

In his widely read book *The Tipping Point* (2000, p. 12), journalist and author Malcolm Gladwell defines the tipping point as "the moment of critical mass, the threshold, the boiling point...that magic moment when an idea, trend or social behavior crosses a threshold, tips, and spreads like wildfire." Depending on the issue, place and timing, what causes a tipping point can range from a new fashion trend to the popularity of a product or a phenomenon such as an increase or drop in the crime rate (ibid.). In 2007, satellite data indicated that Arctic sea ice plummeted to its lowest level since 1979, possibly marking such a tipping point in the way we understand global climate change. In 2012, a new record, surpassing the 2007 level, was set. While we will return to the question of whether the sea-ice minima were portrayed as a tipping point in the media, we could at least say that qualitatively and in general global climate change has morphed into a mediatized *meta-event,* often decorated with polar bear and crumbling sea-ice images, signaling open-ended futures. Yet the well-known ups and downs in the media's 'issue attention cycles' persist and climate change is no exception despite having gained more significance on the public agenda.

With the discursive power of the media in mind, this chapter presents a study of how Arctic climate change has been treated in the quality press between 2003 and 2010. Particular attention has been paid to representations of the 2007 sea-ice minimum, following the release of data from satellite observations, in three daily newspapers: *The New York Times* (United States), *The Guardian* (United Kingdom) and *Dagens Nyheter* (Sweden). The discussion focuses both on the broader frames and issues in the news stories, and on the particular question of a change of media attitude in understanding climate-change indicators such as the changes in the Arctic sea ice.

The power of media institutions to 'steer' political debate and policy has been illustrated in various ways and social contexts (for example the social impact of negative and xenophobic media coverage of migration, unemployment and terrorism in countries such as Denmark, the Netherlands, the United States and the United Kingdom). Accounting for the longitudinal dimensions in the coverage of high-magnitude issues (for example migration and crime) in order to be able to grasp the significance of certain moments in the overall representation of the phenomenon remains important. Climate change is one such high-magnitude issue and the analysis presented here points to the need to understand it as a meta-event with certain moments such as the Arctic sea-ice minima serving as discursive platforms adding new issues and dimensions to the public debate.[1]

Henry Huntington's chapter in this volume tells the media story that *wasn't* and how the 2007 ice minimum was a nonevent to most of the polar indigenous people (especially in Alaska and Russian Chukotka). This chapter offers another account of the shrinkage in Arctic sea ice based upon the ways in which it was framed in international and Swedish national print news. As my analysis of the three newspapers between 2003 and 2010 indicate, two general features characterize the Arctic climate change coverage: *scalar transcendence,* or framing of the issue increasingly as a global or transcendent one rather than a locally or regionally contained phenomenon, and *topical multiplicity,* or the combining of multiple and complex questions in the media stories.

Context and background: media, news and the question of climate change

The ways in which questions of climate change and ecosystems research are taken up within the scientific community and policy circles are marked by a variety of approaches. For the general public, mass media, particularly television and daily newspapers, constitute major sources of information about scientific issues including climate-related questions, although the significance of online sources has increased considerably. As covered extensively in media and communication studies literature, media's role in the coverage of scientific and policy-related issues is far from truly balanced and is governed by various factors, some of which will be discussed shortly. And, as has been documented in an increasing number of publications, media coverage of scientific issues is not a direct reflection of scientific debates and discourses but rather part and parcel of a complex web of power geometries (inter- and intra-institutional); professional/social interactions between policymakers, scientists and media actors; and journalistic norms and routines.

Environmental problems and global warming remain mostly abstract and are not directly and immediately relevant in terms of everyday life and sensory experience (cf. Nelkin, 1995; Antilla, 2010). Thus, news media's framing of issues such as climate change, global warming and international negotiations related with these phenomena have been particularly important in shaping attitudes towards such questions in the public domain. For example, research on the mass media coverage of climate change in the United States and United Kingdom points to the significant role the media played in shaping public opinion and policy decisions. Various studies also highlight a number of factors – from training/educational level of journalists to political economic dictates of media ownership structures – that influence media coverage (Boykoff and Rajan, 2007). As Trumbo and Shanahan (2000, p. 200) suggest, communication plays 'a pivotal role in how governments and societies face this issue and the changes it may bring'.

Further, as Antilla (2010) notes, when a particular issue is covered substantially by the media, its significance (and, the priority attributed to it) in the public eye also increases. An issue or an event is not paid much attention to until it 'reaches saturation coverage and continues to make the news regularly for an extended period of time' (Bennett, 2007, cited in Antilla, 2010, p. 241). Yet whether a particular scientific and political issue or question is elevated to the level of saturation coverage or even a 'media event' is not always reflective of its actual significance relative to other social questions. Likewise the longevity of media attention and news coverage is dependent on factors other than the salience of a given topic as would be deemed by a given expert community or by public opinion. As Djerf-Pierre (2012) notes, issue fatigue and issue competition are two key factors that influence the longitudinal development of issues in the news.

Media's coverage of the human impact of climate change and its role in influencing policy circles and public opinion has mainly been pertinent in the second half of the twentieth century. Particularly from the 1980s onwards, environmental issues gained more visibility in the public sphere through various mechanisms and social movements. Close analyses of media coverage of climate change over the years revealed distinct patterns in journalistic practices. Wilkins and Patterson (1991), for example, suggested that in 1987 and 1988 media frames focusing on scientific aspects were replaced by frames focusing on policy aspects, and that climate-related issues were eventually dropped since they were not linked with any political symbol at the time or they were pushed out as the media tend to prioritize political and economic topics. In her analysis of three decades of Swedish television news, Djef-Pierre (op. cit.) goes on to show how environmental news is crowded out by economic news and news on war and armed conflicts in times of crises. Discursive shifts in the coverage of environmental and climate change are also clearly discernable. As Carvalho (2005) illustrates, in the period preceding 1988, 'greenhouse effect' was the dominant term used in the media to denote climate change. Greenhouse effect was succeeded by 'global warming' in the following years. In 1990, it became the most frequently used phrase in the media in reference to climate change (ibid.).

Different terminology carries (radically) different connotations in different contexts. Although 'climate change' has been a common topic of public debate in international media as well as in the United States, as noted by Gelbspan (2004, p. 67), American media have 'consistently minimized this story'. It is also easily deducible from the American political discourse that while 'climate change' is perceived as more politically neutral, 'global warming' is seen as more alarmist and value-laden. What is clear today, compared to the preceding decades, is that climate change and the consequences ensuing from it have also been cast as political and social, not merely scientific, issues posing a challenge to both governments and citizens (cf. Carvalho and Peterson, 2009). In its 2007 evaluation report, the Intergovernmental Panel on Climate Change (IPCC) concluded that surface temperatures were on the rise, and 'climate change was used to refer exclusively to changes in the climate caused by human activities:

> it is extremely unlikely that global climate change of the past 50 years can be explained without external forcing and very likely that it is not due to known natural causes alone (IPCC, 2007, p. 39)

As Russill and Nyssa (2009) discuss, in the United States the Republican Party adopted the term 'climate change' as a strategic move in 2002, suggesting reference to general change in the climate due to a variety of causes, while global warming is more specifically associated with human influence. Russill and Nyssa (op. cit.) further note, in reference to Boykoff's analysis (2007) that newspaper coverage of climate change in US prestige press increased about two and a half times between 2003 and 2006. Further, American network television evening news coverage of climate change increased from less than 10 news segments in 1995 to over 20 in 2004, with over 40 segments in 1997 (Boykoff, 2008,cited in Russill and Nyssa, 2009). As demonstrated in the results and discussion part of this chapter, there has also been a general increase in the volume of news reports covering climate change over the period from 2003 to 2010, with 2010 actually marking an overall low point in the coverage of climate change by American mass media.

Despite the increasing significance of climate change, various studies confirm that the coverage of climate change has had an episodic character (cf. Boykoff and Boykoff, 2007; Hutchison, 2008). In episodic coverage (as opposed to thematic), the mediated debate about an issue materialized around events. The coverage of the 2009 Copenhagen climate summit (COP15) constitutes one such event (see Christensen, 2011). Going further back, there was a major drought in 1988; 1997 was the year of the Kyoto Agreement; in 2005 Hurricane Katrina hit New Orleans; in 2007 Al Gore's documentary *An Inconvenient Truth* was released; and also in 2007 Al Gore and IPCC were awarded the Nobel Prize (see also Boykoff, 2008b). However, significant events (such as US election campaigns and different views of candidates on climate issues) other than the Arctic sea-ice retreat play a role in this increase during certain moments and periods. It could be suggested that the sea-ice retreat of 2007 was among the significant events that the media were able to link with climate change on the whole. This also taps into the question of whether the shrinking Arctic sea ice constituted a tipping point in the mediated representation of climate change.

Holland et al. (2006), in referring to the changes in the Arctic sea ice, pose the question of whether a tipping point has been reached in climate change and how best to assess it. As Russill and Nyssa (op. cit.) note, 'discussions of tipping points in climate science often point directly to the popular influence of Malcolm Gladwell'. As Gladwell (1996; 2000, p. 9) describes it, the tipping point is 'that one dramatic moment in an epidemic when everything can change all at once...'. What characterized a tipping point is 'sudden change', not 'steady progression' or 'proportionality' (pp. 12–13). Hence, communication (and repetition and narrative) remains vital for Gladwell in impacting public perception. We will return to the question of tipping point while discussing study findings.[2]

Media, public sphere and framing

Frame analysis is widely used in media studies to analyze both the content and the impact/influence of news coverage. While Goffman's work (1974) marks a milestone in establishing frame analysis as an instrumental method in content analysis in general, current usage of frame analysis is largely based on more recent work on framing in news, where many studies have employed the concepts of media frames and framing in news (for example, Altheide, 1997; D'Angelo, 2002; De Vreese et al., 2001; Entman, 1991, 1993, 2004; Reese and Buckalew, 1995; Scheufele, 1999, 2000; Semetko and Valkenburg, 2000). A media frame can be described as 'a central organizing idea or story line that provides meaning to an unfolding strip of events.... The frame suggests what the controversy is about, the essence of the issue,' Gamson and Modigliani (1987, p. 143). The *absence* of certain aspects and information in a news story is as crucial as what is included in shaping meaning:

> Frames are the focus, a parameter or boundary, for discussing a particular event. Frames focus on what will be discussed, how it will be discussed, and, above all, how it will not be discussed. It is helpful to think of 'frames' as very broad thematic emphases or definitions of a report, like the border around a picture, that separates it from the wall, and from other possibilities. (Altheide, 1997, p. 652)

A significant volume of research has been published in recent years on the coverage and representation of climate change in the (mainly Western) media (for example, Carvalho, 2010; Anderson, 2009; Boykoff and Roberts, 2008; Boykoff, 2007; Eskjaer, 2009; Olausson, 2009; Nerlich et al., 2010; Nisbet, 2009; Boykoff and Boykoff, 2004; Boykoff, 2008a, 2008b; Lyytimäk and Tapio, 2009). These works have at their core the proposition that climate change is a complex process, and that the media play a central (albeit often flawed) role in relaying information regarding this process to the general public. As Anderson (2009, p. 166) writes,

> The media play a crucial role in framing the scientific, economic, social and political dimensions through giving voice to some

viewpoints while suppressing others, and legitimating certain truth-claims as reasonable and credible. While media have been shown to play a key role in framing climate change, the effects are complex and dynamic and there is no straightforward relationship between information campaigns and behaviour change.

One of the central issues for research into the framing of climate change in the media is the extent to which the media both reflect and influence the science and policies related to climate change. In their study, Boykoff and Roberts (2008) noted that while it was clear that science can influence journalism and the public understanding of science, at the same time journalism (and other media coverage) can also influence science, and the policies related to science and climate change. In the end, the authors' note, the line between reporting on science and attempting to 'educate' the population becomes blurred. As Carvalho (2010, p.172) notes, citizens access scientific information and political debates primarily through the media discourse they consume:

> Media coverage has been a key factor in raising levels of awareness and concern in the last decade or so. Representations of the problem in the media are also likely to have influenced citizens' understanding of both the risks associated with climate change and the responsibilities in addressing the problem. Moreover, the social credibility and social authority of different social actors, their claims and arguments, are also largely defined by discursive exchanges taking place in the media.

Finally, in regard to journalistic coverage of climate change in general, Trumbo and Shanahan (2000) surmise that the vociferous debate over climate change that occurred in the late 1980s was a battle waged through communication and miscommunication. Thus to fully understand the role of the media in influencing public agendas, construction of narratives also emerges as a key area that needs close analysis. Jacobs (1996) suggests that 'News work is not merely an instrumental task of "filling the news hole"; [it] requires the transformation of discrete events into meaningful narratives' (cited in Trumbo and Shanahan, 2000, p. 201).

Methodology

This study is divided into two parts: a preliminary quantitative account of articles in the three newspapers in question (*The New York Times*, *The*

Guardian and *Dagens Nyheter*) designed to indicate both longer-term trends in terms of subjects covered and volume, followed by a qualitative frame analysis of a selection of these articles. The selection of newspapers is based on the assumption that such major media outlets are indicators of larger media trends (Boykoff, 2009). The newspapers chosen are *Dagens Nyheter* (Sweden), *The Guardian* (United Kingdom) and *The New York Times* (United States). The purpose of including newspapers from three different countries (representing three different political and cultural climates) is to produce more comprehensive and comparative knowledge on a question with global significance. The three newspapers belong in the broad spectrum of what is labeled 'quality press' and 'prestige/elite press'. Therefore they do not span the political spectrum from right to left.

In order to locate relevant articles, a search was conducted on the websites of all three newspapers. First, a search was done for articles including the term 'climate change' between the dates of 1 January 2003 and 31 December 2010 (inclusive). The time period covers roughly the four years preceding and four years following the 2007 satellite observations of the Arctic sea-ice minimum.[3] 2003 is also the year when the decline in ice in the Arctic area started to accelerate sharply. It is important to note that the year 2001 was also significant in the climate-change debate. The climate-change issue (including the earlier conceptualization terms such as 'greenhouse effect') gained political significance in the British press in 1985, with other milestone moments and discursive shifts in the late 1980s and throughout the 1990s, as well as a decline between 1991 and 1996. In 2001, the release of IPCC's Third Assessment Report linked climate change with human impact, which caused a major jump in the news stories addressing the issue, and the coverage increased steadily from that point onwards (Carvalho, 2007, p. 226).

All 'hits' in the searches conducted for the study presented here were included in the sample size (regardless of the section of the newspaper within which the article was published). From this base sample of stories, further search terms were successively added in order to ensure that the articles selected addressed: (1) the question of 'climate change' in relation to the 'Arctic' region; and, (2) as a different category (as will be discussed shortly), 'climate change' and the 'Arctic' in relation to sea ice (and ice retreat). (The results of this section of the study are outlined in Section 5.)

A qualitative frame analysis follows the preliminary quantitative analysis, based on the newspaper pieces categorized as world/international and domestic *news* in the online archives and gathered by utilizing the keywords 'climate change' AND 'Arctic' AND 'sea ice'. The qualitative content analysis involves frame analysis as an instrumental method to highlight the central organizing ideas and story lines in the material covered. Among the elements that were considered in the analysis were **title of story**; **topical/thematic construction** (that is, the dominant themes and points that guide the story); **incidence/contextual frame** (in other words, is the issue invoked in relation to a specific event/moment or in general); **narrative structure**; **voice/lack of voice** (that is, whose concerns are voiced from whose perspective in reference to knowledge/facts/claims provided by whom; and who is left out); **explicit points/suggestions and implications**; and, **choice of terms**. While the entire textual body of each piece was included in the analysis, particular attention was paid to the opening and closing paragraphs (as they carry significant weight in journalistic coverage), to parts that clearly stood out in the overall narrative, and to quotations when and where applicable.

Results of the quantitative analysis and selection criteria for the qualitative analysis

The breakdown of the number of articles is presented in Tables 2.1 and 2.2.

As the tables illustrate, there were a great many pieces that appeared, including the phrase 'climate change' in both the *New York Times* and *Guardian* between 2003 and 2010, with a significantly lower number published in *Dagens Nyheter* during the corresponding period. However, the total number of hits decreased dramatically when the term 'Arctic' was added to the search. Only 7 percent of the *New York Times* pieces between 2003 and 2010 that contained the term 'climate change' also contained the term 'Arctic', with only 4 percent of *Guardian* hits doing the same (although the overall number of articles in *The Guardian* is far larger).

The research project leading up to this volume had a focus on the impacts of climate models for the public understanding of climate change. From this perspective, it is interesting to briefly note in passing here that results of the broad survey of articles indicated that fewer than 1 percent of all pieces published between 2003 and 2010 containing the term 'climate change' made any reference to a model, models, or modeling (although these figures are not included in the Tables 2.1 and 2.2).

One of the questions raised in the research process was whether the 2007 Arctic sea-ice minimum played a major role in the mediated

Table 2.1 News pieces published 2003–2006

	Containing term 'climate change'	Containing terms 'climate change' AND 'Arctic'
The Guardian	5583	267 (5%)
The New York Times	801	81 (10%)
Dagens Nyheter	44	3 (7%)
Total	6428	351 (5%)

Table 2.2 News pieces published 2007–2010

	Containing term 'climate change' (and % increase from 2003–2006)	Containing terms 'climate change' AND 'Arctic'
The Guardian	16,134 (+188%)	671 (4%)
The New York Times	4,364 (+445%)	265 (6%)
Dagens Nyheter	71 (+61%)	6 (8%)
Total	20,569 (+220%)	942 (5%)

discourse about climate change. Two assumptions would appear to be logical in this case: (1) that the overall number of stories on climate change would be larger in 2007–2010 than it was in 2003–2006; and (2) that stories also including the term 'Arctic' would increase not only in volume (a logical assumption if one assumed that the overall numbers would increase), but also in proportional terms, with a higher percentage of stories including the term in 2007–2010 as compared to 2003–2006. In order to see the overall change in the framing and discursive construction of Arctic climate change, the newspaper articles were broken down into two halves: data from 2003–2006 and from 2007–2010.

As tables 2.1 and 2.2 indicate, the assumption that the overall number of pieces on climate change would increase was proved correct. A dramatic increase in the number of hits containing the term was seen in 2007-2010, with an overall threefold increase of the total number of hits (from 6,429 to 20,582). This significant jump in the number of newspaper pieces containing the term 'climate change', however, was not matched by an increase in the proportion of articles addressing the Arctic. In fact, although there *were* more articles written on these issues after 2007, proportionally speaking both the *Guardian* (5 percent versus 4 percent)

Table 2.3 News pieces published 2003–2010

	Containing term 'climate change'	Containing terms 'climate change' AND 'Arctic'	Containing terms 'climate change' AND 'Arctic' AND 'sea ice'
The Guardian	21,717	938 (4%)	242 (1% / 26%)
The New York Times	5,165	346 (7%)	107 (2% / 31%)
Dagens Nyheter	115	9 (8%)	4 (3% / 44%)
Total	26,997	1293 (5%)	353 (1% / 27%)

and *The New York Times* (10 percent versus 6 percent) were slightly *less* likely to include the terms 'Arctic' when discussing climate change than they were before 2007. In order to achieve more meaningful and representative results and get a broader spectrum of the news frames that were utilized in the three newspapers, a second round of archival search, this time using 'sea ice' as the third set of keywords was conducted.

As Table 2.3 shows, the search combining this particular combination of keywords yielded a significantly higher number of stories in the third category. Of the pieces containing *both* climate change and Arctic, 27 percent made reference to sea ice (while eliminating the term Arctic yielded a proportion of 1 percent, meaning only 1 percent of news stories related *only* with climate change *also* made reference to sea ice). A general scanning of the newspaper pieces in this particular group also indicated that clearer patterns and frames were discernable as to how the question of Arctic climate change was taken up during this eight-year period. Considering the large number of articles, and, in order to be able to conduct a content analysis based on a close reading of the pieces, further search criteria were introduced. While many of the newspaper stories appeared in editorials, opinion pieces and sections ranging from business/economy/finance to culture to obituaries and book reviews, through a last round of sifting and sorting of the search results, only those that appeared in environmental/science and news (domestic and world) sections were listed as the Tables 2.4–2.6 illustrate.

Please note that in *DN*, hits containing only the terms Arctic (*Arktis*) and sea ice (*havsis*), and without the term climate change (*klimatförändring*), totaled much higher with 30 articles between 2003 and 2010. However, for consistency, only those containing the three sets of keywords (as in the case of *The New York Times* and *The Guardian*) were taken into consideration for the purposes of the discussion here.

Table 2.4 Sectional breakdown of the pieces containing climate change, the Arctic, and sea ice published in *The Guardian* in the period 2003–2010

The Guardian	Environment-Science	World News	UK News/News (inclusive of blog pieces)
2003–2006	24	3	4 (incl. 1 blog, 1 weekly summary series entitled The Wrap)
2007–2010	146	9 (incl. 1 blog piece)	4 (incl. 3 blog piece)
Total	170	12	8

Table 2.5 Sectional breakdown of the pieces containing climate change, the Arctic, and sea ice published in *The New York Times* in the period 2003–2010

NYT	Science	World News	US News/News
2003–2006	9	3	1
2007–2010	9	3	7
Total	18	6	8

Table 2.6 Sectional breakdown of the pieces containing climate change, the Arctic, and sea ice published in *Dagens Nyheter* in the period 2003–2010

DN	Science	News
2003–2006	NA	1
2007–2010	NA	1
Total	–	2

The following section offers a qualitative analysis in order locate the thematic frames used to mediate climate change, the Arctic and sea ice, and to do so with a particular eye towards any shifts that took place in the nature of the coverage after 2007 (when the sea-ice satellite data entered into the scientific and popular arenas). Considering that there is a large number of total pieces in the three sections combined (science/environment; world/international news; and news/domestic news), with many of those appearing in the science/environment sections as special stories/reports of significant length, and in order to be able to delineate a meaningful analytical category, only pieces that appeared in the 'news' (world and domestic combined) sections of the three newspapers were included in the final qualitative analysis. It is in line with previous research (for example Olausson, 2009) to select a limited number of categories, or a single category, for a more detailed qualitative analysis. In the case of the current study, the 'news' category is optimal since it represented the widest variety of stories on climate change.

Results of the qualitative frame analysis: ice, water and bears

The qualitative analysis of the selection of articles from *Dagens Nyheter*, *The Guardian* and *The New York Times* examines how the issue of climate change was represented and discussed in relation to the Arctic region and receding sea ice. In identifying the broader frames, earlier studies on the media coverage of climate change were also consulted. Boykoff (2008c, p. 555) notes that 'media framing involves an inevitable series of choices to cover certain events within a larger current of dynamic activities'. Bennett (2002, p. 42, cited in Boykoff, op. cit.) refers to the process of media framing as 'choosing a broad organizing theme for selecting, emphasizing and linking the elements of a story such as the scenes, the characters, their actions, and supporting documentation'. Boykoff (op. cit., pp. 555–6) uses the following broad categories where subcategories are included in order to illustrate the types of subject matter that fall under each larger frame:

Ecological
- Weather events (for example heat waves, droughts, floods)
- Biodiversity (for example plants, animals)

Political and economic
- Political actors (for example UN meetings; rhetoric, action)
- Economics and business

Culture and society
- Popular culture (for example celebrity movements, royal family, films and books)
- Justice and risk, public health (ethics, inequality, adaptation)
- Transport
- Public understanding, knowledge, education (for example, poll results, consumer reports)

Scientific

- Discoveries, fundamentals, new studies
- Science funding and processes
- Applied science, technologies (for example renewables)

General

- Other

(Boykoff, op. cit., p. 556)

Boykoff's broad categories were unmistakably applicable and present in the pieces analyzed in this study (Table 2.7).

While the results of this study reveal that the discursive constellation of climate change + Arctic + sea ice was framed heavily in relation to political, economic and scientific issues, in many cases cross-bridges were built between two or more frames and the melting ice in the Arctic and its consequences were addressed as both a global and local risk category (what I call **scalar transcendence**) and in relation to multiple complex questions such as political action, scientific un/certainty and wildlife (**topical multiplicity**).

Although Boykoff identifies one more frame in his study in conjunction with 'ecological' ('meteorological', such as the reporting of weather events, droughts and so on), this was not found to be a prominent frame in this analysis. If weather events such as the flooding of coastal areas in Alaska, for instance, were noted, it was done so in relation to political frames (for example the town's filing a lawsuit) and/or cultural/social frames and scientific certainty based on meteorological events providing further proof for global warming. The ecological frame was an important one, particularly in relation to local communities and polar bears, and, of the 36 total stories closely analyzed, nine were about/or made mention of polar bears – but mostly in relation to political/legal action. In addition to the main frames, Table 2.7 lists the subframes and topical areas as covered in the collection of newspaper pieces analyzed. In what follows, I offer a discussion of these frames identified in the study: scientific; political and economic; culture and society; ecological; and other.

Scientific frame

A major issue in the mediated discussions about climate change has been the question of scientific 'certainty' of the prediction of climate change (for example whether there is sufficient certainty to justify political action). One of the notable findings here (based on the sample pieces) is that the frame of scientific 'certainty' and related discussion was apparent in the media coverage of scientific facts and projections about the shrinking ice in the Arctic as early as 2003 – the start of the archival search period as identified in this study – in *The New York Times* and *The Guardian*. To begin with, the *DN* piece entitled 'An Ice-free Arctic is Completely Possible' (2006), referred to scientist Erland Källén from Stockholm University who suggested that the Arctic ice carries great significance in relation to how climate develops and changes. Another *DN* article from 1 February 2007 ('Greenlandic Ice is Melting Faster than Anyone Thought') is exemplary of the multiplicity of issues covered in relation to melting ice and how the newfound scientific certainty (as indicated by various new research) as to this phenomenon served as a catalyst to bring public attention to other political, social and moral issues related with global climate change (that is, temperature rise; implications for humanity; need for Swedes and Sweden to take the lead and take action rather than just focus on their own footprint; Swedish prime minister Reinfeldt's pointing a finger to gas emissions of India and China; EU decisions; and other questions materializing around such core ones).

Most stories included in the sample group were about how fast the ice had been melting, its global/local implications and how the Arctic ice shrinkage had been underestimated. This was by far the most prominent theme (covered under all four categories of the main frames) that guided the news articles before and after the 2007 sea-ice minimum. Coverage of the shrinking ice in the Arctic often made reference to data retrieved from satellite scanning of the area, both before and after the 2007 sea-ice minimum. While the difference between what has become clear (that is a rapidly melting ice) at the current stage and what was estimated before (that is a slower ice retreat) was a predominant entry point and narrative element in many of the stories, this was not attributed to the discrepancies between the modeling data and satellite data. In fact, to a lay person not aware of the scientific differences between climate modeling and satellite observations (and debates around these), the news stories would not indicate a disjuncture between the two. This is largely due to the fact that most stories noting an unexpectedly fast melt-down did so by way of using comparative expressions such as 'more/or faster than expected before' and its variants. For instance, a *New York Times* story entitled 'Arctic Sea Ice Melting Faster, a Study Finds' (1 May 2007) notes 'Climate scientists may have significantly underestimated the power of global warming from human-generated heat-trapping gases to shrink the cap of sea ice floating on the Arctic Ocean, according to a new study of polar trends', citing a study published in *Geophysical Research Letters*.

A news story ('The Arctic: here today gone in 40 years', 2006, *The Guardian*) refers to a 10-second animation 'providing a scary illustration of the possible impact of climate change', produced by the National Centre for Atmospheric Research in Colorado. Another one, from *The Guardian*, entitled 'Meltdown fear as Arctic ice cover falls to record winter low' (15 May 2006) noted the 'alarming' decline in the Arctic ice volume. In reference to Walt Meier from the US National Snow and Ice Data Center in Colorado, the story raised the question of an Arctic tipping point:

> Experts are worried because a long-term slow decline of ice around the north pole seems to have sharply accelerated since 2003 raising fears that the region may have passed one of the 'tipping points' in global warming. In this scenario, warmer weather melts ice and drives temperatures higher because the dark water beneath absorbs more of the sun's radiation. This could make global warming quickly run out of control.

It further quoted Meier who suggested, 'For 800,000 to a million years, at least some of the Arctic has been covered by ice throughout the year. That's an indication that, if we are heading for an ice-free Arctic, it's a really dramatic change and something that is unprecedented almost within the entire record of human species.' While this piece was one of the two stories that made explicit mention of a 'tipping point' in climate change, various other stories raised the issue using other terminology (such as 'alarming' change) and narrative strategies to indicate that the melting Arctic sea ice is indeed to be seen as symptomatic of a 'point of no return' in global climate change. Among the other scientific issues noted in the overall body of the news reports were the presumed positive feedback loop to kick in by 2025 (creating a vicious circle and more surface warming); other 'alarming' signs such mosquitoes and 'freakishly' high temperatures; calculations of how much emission cuts are needed to avoid the 'Armageddon' (noting the 'disappearing world of the Arctic'); measurements of glaciers and sea-level rise; consensus among scientists (for example, 'World Scientists Near Consensus on Warming', *New York Times*, 30 January 2007); and, to a much lesser degree, controversy among scientists (for example Jack Krupansky from Washington claiming, 'It is simply not credible for any of these scientists to "suggest" that' climate effects will carry into the future in a 2006 *Guardian* piece). It should be noted that 'uncertainty', in this body of articles, was not a prominent theme and was almost exclusively mentioned in relation to climate-skeptic American politicians such as Palin and Bush (other than a very few cases).

Political and economic frames

Analysis of the political frames revealed that emphasis was placed on both the need for and lack of political action and mitigation. Responsibility of the big polluters such the United States, China and India was pointed to in a number of the articles. As discussed earlier, certain 'events' and social/political/scientific moments of significance are what drive and boost media coverage in relation to issues such as climate change. This trend was most apparent when analyzing the political and economic frames. As listed in Table 2.7, American politics was heavily covered. A *Guardian* piece from 21 December 2008, 'Obama's revolution on climate change', read:

> Barack Obama ushered in a revolution in America's response to global warming yesterday when he appointed one of the world's leading climate change experts as his administration's chief scientist. The president-elect's decision ... reveals a new determination to draw a line under eight years of US policy that have seen George Bush steadfastly reject overwhelming evidence of climate change.

Similar stories appeared in *The New York Times*. The Bush administration's skeptical attitude towards climate change and its resistance to curbing emissions were also among the political topics noted. A prominent topic, as it appeared in this sample of articles, was Sarah Palin's war against the polar bear, producing headlines such as 'Sarah Palin v the polar bear: who will survive?' (2 October 2008, *Guardian*); 'Palin fought safeguards for polar bears with studies by climate change sceptics' (30 September 2008, *Guardian*); 'Revealed: oil-funded research in Palin's campaign against protection for polar bear' (1 October 2008, *Guardian*). Interestingly, Palin was a more prominent story in *The Guardian* in relation to the three search terms. While the story was covered by *The New York Times* as well, those stories did not feature among the pieces that were collected as hits after the keyword search. Other political issues that commonly appeared in news stories were the question of legal protection for polar bears; Russia's territorial claims to the Arctic; and the ban the Alaskan division of the Federal Fish and Wildlife Service imposed on using certain terms by its scientists and employees ('Memos Tell Officials How to Discuss Climate', *New York Times*, March 8, 2007):

> Internal memorandums circulated in the Alaskan division of the Federal Fish and Wildlife Service appear to require government biologists or other employees traveling in countries around the Arctic not

to discuss climate change, polar bears or sea ice if they are not designated to do so.

New opportunities for oil-drilling and related perils; Russia's plans to build nuclear stations to exploit Arctic oil; and increasing opportunities for cod fisheries and economic returns (which would, ironically, lead to more environmental destruction and warming as the same stories remarked) were among the issues covered both in relation to global consequences and contingencies in their local contexts. This is illustrative of scalar transcendence in the news stories included in this study. Not only are the ways in which local and global dynamics influence each other noted, but the inherently uncertain nature of 'environmental destruction' in general (that is, where it can hit and its extent) invokes the transcendent (or all-encompassing) spatiality of Arctic climate change.

Culture and society frame

One of the changes observed in the coverage of climate change in the media in general has been the growing significance of cultural and societal frames, and political frames. While, earlier, climate change was invoked mostly as a purely scientific issue, there has been a discursive shift, particularly in response to the mounting evidence for global warming, towards privileging societal and political dimensions. As noted in Table 2.7, the topics that appeared under this category included disappearing local languages and cultures in the northernmost regions; flooding of coastal villages; the question of local livelihood and hunting bans; moral and ethical responsibility of the developed nations; and communal action. As in the other cases, spatial dynamics were evident here with such issues being brought up both in terms of their global/transnational relevance and local significance/immediacy.

A *Guardian* article dated 28 November 2010, 'In a far corner of Greenland, hope is fading with the language and sea ice', climate change, hunting controls and a new consumerism were highlighted as factors threatening the way of life of the Polar Eskimos. The piece notes that 'the climate in north-west Greenland has become very unpredictable in recent years and hunters no longer know when the sea ice will come or how long it will stay for'. Ethical and moral dimensions of the global dominance of Western nations and cultures are also highlighted in the story:

> In this closed, inward-looking society, the Polar Eskimos are sometimes wary of the outsider and the finger of blame is often pointed at the white man whose market capitalism, individualism and climate

change are perceived as catalysts in the demise of their own traditional communal group culture, damaging social cohesion in the process. With the recent EU ban on all seal products, this is even more the case now that the export market for their goods has completely disappeared.

Another story from the same newspaper ('The disappearing world of the last of the Arctic hunters', 3 October 2010) remarks that 'Global warming has a human cost too, tearing families apart. To visit their Canadian relatives, these people would now have to fly to Copenhagen 4,000 km away then across the Atlantic to Montreal and up from there.' A *New York Times* piece from 15 December 2004 ('Eskimos Seek to Recast Global Warming as a Rights Issue') about Eskimos' seeking a ruling from the Inter-American Commission on Human Rights regarding the United States on the issue of human rights abuses further reports that

> Last month, an assessment of Arctic climate change by 300 scientists for the eight countries with Arctic territory, including the United States, concluded that 'human influences' are now the dominant factor.

The same article also includes a quote: 'Something is bound to give, and it's starting to give in the Arctic, and we're giving that early warning signal to the rest of the world.' The issues the story covers and the narrative structure combine both the global and the local and links the scientific frame (certainty) with societal ones such as civic action, moral/ethical responsibility and public safety.

Ecological frame

In the coverage of ecological questions in relation to the Arctic and sea ice, the polar bear featured as a regular theme (and a charming visual character deliberately used) in a number of the stories. Among the issues brought up with regularity was the disappearing habitat of the bears (and, but not so much, of other Arctic animals) and their status (endangered or not), which was invoked mostly in relation to legal/political issues as discussed in the relevant section above. Increased cod fisheries in southern Greenland were discussed in relation to both economic and environmental consequences. This frame, in and of itself, was not an autonomous or highly prominent one and it was (in each and every case) covered in relation to one or more of the other frames.

Final thoughts

As Rick, Boykoff and Pielke (2011) remark, there are a number of prominent voices in the scientific community, such as Dr. James Hansen of NASA GISS, who advocate for the use of stronger language when discussing climate change in scientific literature and criticize glaciologists for shying away from expressions of urgency when publishing on, for instance, sea-level rise. As the media (often sensationally) cover, there are disagreements in the scientific community concerning the ways in which climate change in general and the Arctic sea-ice retreat and sea-level rise in particular should be communicated to the public. The media story that *was*, as discussed above, is based on news stories from three newspapers with many other articles and news outlets having been inevitably left out due to the limited scope of this study. Thus, while it is certainly impossible to reach broad conclusions – without them being speculative – as to the mediation of climate change in general and Arctic change and sea ice in particular, certain patterns and discursive trends are clearly discernible based on the sample group analyzed here. It is worth mentioning that this chapter is an effort to examine relatively longer-term discursive elements surrounding the Arctic and climate change, rather than a focus on short-term dynamics, and it would not be far-fetched to suggest that these news pieces do have a certain degree of representative power in talking about the recent, qualitative trends in the quality press coverage of the questions identified in this chapter.

We can then conclude our discussion of these media stories with a few humble remarks. The analysis here (particularly in comparison to earlier studies on similar questions) reveals that there is a mediated return to science and scientific truth, this time with a positive tone about the success of scientific quests, which proved that the planet is indeed warming. Unlike the earlier news stories from the 1980s and portions of the 1990s which 1) placed a strong emphasis on scientific controversy and uncertainty; and 2) cast 'climate change' in scientific isolation (that is, a battle fought among science geeks) rather than its sociopolitical significance, the growing scientific certainty, from 2003 onwards, about Arctic climate change as bellwether for global warming was linked with sociocultural, economic and political frames. Stories that brought in the uncertainty factor were few and far between (and mostly linked with the extreme politics of those such as Palin and Bush, de facto undermining the question of uncertainty).[4]

Differences between climate models and satellite data/other observations were not among the explicit debates. In discussing scientific

Table 2.7 Frames, subframes and topics

I. Political Economic	II. Scientific	III. Culture and Society	IV. Ecological	V. Other
1) Politicians/Social Actors • Bush's climate policy • International regimes and agreements • Responsibility of big polluters (US, China, India) • McCain's pro-action politics about global warming • Obama's appointment of J. Holdren • Palin's fight against global warming (and polar bears) • Censorship on Alaskan Div. of Fed. Fish and Wildlife Service to discuss polar bears, sea ice and climate change with locals in the Arctic region • Legal protection of polar bears • Room for action for smaller players (e.g. Sweden) *2) Business and Economy* • Environmental impact of Greenland's economic development	*1) Certainty* • Arctic as barometer • Ice retreat (both Arctic and in some articles, also Antarctic) • Greenland ice cap • Rising atmospheric and sea temperatures • Rising sea levels • Change in ocean currents • Significance of satellite data (particularly after 2007) • Human impact • Models confirming satellite-based predictions • Consensus among world scientists	*1) Local culture and Languages* • Arctic communities moving away and dying away of languages and local customs. *2) Communal Livelihood* • Positive impact: global warming and more livestock in southern Greenland (e.g. increased cod fisheries) • Negative impact: polar Eskimos in Northwest Greenland; hunting control and warming (i.e. disappearing of exports and unpredictability of ice come-and-go); hunters falling into thinned ice; disappearing of transport routes b/w Canada and Greenland	*1) Wild life* • Polar bears (food supplies, disappearing habitat) • Protection of polar bears • Controversy around hunting restrictions • Disappearing natural environment	• Armageddon and the last decade to save the planet • Spiritual leaders visiting the glaciers and praying • Need for icebreaker ships

- Russian plans and oil reserves in the Arctic
- Russia's ownership claim to North Pole area
- Pros/cons of open-sea shipping in now-ice-free areas

2) Conflict and Uncertainty
- Variability
- Credibility of methods and scientists
- Climate scientists fight back against sceptics

2) Civic Action
- Inuit seal-hunting communities seeking ruling from the Inter-American Commission on Human Rights that the US, by contributing to global warming, is threatening their existence.
- Relocating in Alaskan village due to flooding and communal lawsuit against corporate business.

3) Morality and Ethics
- Responsibility of the developed West to cut down emissions and take action
- Responsibility to preserve local cultures

5) Public Safety and Hazards
- Polar bears coming into closer contact with people
- Flooding hazard in local coastal towns

dimensions, most stories made reference to American and British (and, in the case of *DN*, Swedish) researchers and research institutes and often cited specific scientific studies. Considering the overall tone, it is possible to suggest that an emphasis on science and scientific truth (that climate change is a reality) was discernible in most stories. The quote below from President Obama ('Obama's revolution on climate change', 21 December 2008, *Guardian*) provides a good example:

> 'Today, more than ever before, science holds the key to our survival as a planet and our security and prosperity as a nation,' Obama announced. 'It's time we once again put science at the top of our agenda and...worked to restore America's place as the world leader in science and technology.'

The issues at stake were covered with respect to their complexity and transgeographic significance. Scientific frames that sent a strong message often provided entry points in terms of building linkages between environmental change and political and social dimensions.

Finally, while Arctic climate change was not cast as a 'tipping point' (explicitly in those terms) save for two examples, it is clear from the narrative buildup and terminological character of the majority of the pieces (for example expressions such as 'a profound transformation of the planet') that the accelerated sea-ice retreat was framed as that 'magic moment when an idea, trend or social behavior crosses a threshold, tips, and spreads like wildfire' (Gladwell, 2000). The gradual shrinkage of the polar ice has provided a new and important gateway to understand the meta-event of global climate change in the concrete and tangible context of the Arctic region.

As we discuss in the final chapter of this volume, the general trend of peaks and dips in the overall coverage of climate change continues, with the issue almost falling off the map in the United States in 2010. While such shrinkage in volume is no small feat and has significant implications for issue-visibility in the public domain, online media and alternative platforms have also become increasingly important to the public for access to information. In qualitative terms, news media have improved their reporting on the issue with environmental journalists particularly grasping and conveying the significance of climatic change. This should give us hope. Future studies will reveal how the media story of the Arctic is being told.

Notes

1. A conceptual discussion of media events (and meta-events) is offered in the introductory essay.
2. See also Nuttall (2012) for a review of tipping points in the social sciences.
3. The satellite data were made public in September 2007. Therefore, in actual terms, the period preceding, as included in this study, is around four years and eight-and-a-half months, while the period following is three years and three-and-a-half months. However, for the sake of a more clear-cut periodic categorization, full four years were used to divide the news stories into two groups (i.e., 2003–2006 and 2007–2010).
4. This, of course, does not negate the widespread discussion around scientific controversy at the time of "Climate-Gate" (of 2009).

References

Altheide, D. L. (1997) 'The News Media, The Problem Frame, and the Production of Fear', *The Sociological Quarterly*, 38, 646–68.

Anderson, A. (2009) 'Media, Politics and Climate Change: Towards a New Research Agenda', *Sociology Compass*, 3:2, 166–82.

Antilla, L. (2010) Self-censorship and Science: A Geographical Review of Media Coverage of Climate Tipping Points. Public Understanding of Science, 19:2, 240–256.

Bourdieu, P. (1977) *Outline of a Theory of Practice* (Cambridge: Cambridge University Press).

Boykoff, M. T. (2007) 'From Convergence to Contention: United States Mass Media Representations of Anthropogenic Climate Change Science', *Transactions of the Institute of British Geographers*, 32, 477–89.

—— (2008a). 'Lost in Translation? United States Television News Coverage of Anthropogenic Climate Change, 1995–2004', *Climatic Change*, 86:1–2, 1–11.

—— (2008b) 'Media and Scientific Communication: A Case of Climate Change', in D. Liverman, C. Pereira and B. Marker (eds), *Communicating Environmental Geoscience*, Geological Society (London, Special Publications, 305, 11–18).

—— (2008c) 'The Cultural Politics of Climate Change Discourse in UK Tabloids', *Political Geography*, 27, 549–69.

—— (2009) '"We Speak for the Trees": Media Reporting on the Environment', *Annual Review of Environment and Resources*, 34, 431–58.

Boykoff, M. T. and J. Boykoff (2004) 'Balance as Bias: Global Warming and the US Prestige Press', *Global Environmental Change*, 14, 125–36.

—— (2007) 'Climate Change and Journalistic Norms: A Case-Study of U.S. Mass-Media Coverage', *Geoforum*, 38:6, 1190–1204.

Boykoff, M. T. and S. R. Rajan (2007) 'Signals and Noise: Massmedia Coverage of Climate Change in the USA and the UK', *European Molecular Biology Organization Reports*, 8, 1–5.

Boykoff, M. T. and Roberts, J.T. (2007/8) 'Media Coverage of Climate Change: Current Trends, Strengths, Weaknesses', *Human Development Report 2007/2008*.

Carvalho, A. (2005) 'Representing the Politics of the Greenhouse Effect', *Critical Discourse Studies*, 2:1, 1–29.

—— (2007) 'Ideological Cultures and Media Discourses on Scientific Knowledge: Re-reading News on Climate Change', *Public Understanding of Science*, 16:223–43.

—— (2010) 'Media(ted) Discourses and Climate Change: A Focus on Political Subjectivity and (dis)Engagement', *WIREs Climate Change*, 1:172–9.

Carvalho, A. and T. R. Peterson (2009) 'Discursive Constructions of Climate Change: Practices of Encoding and Decoding', *Environmental Communication*, 3:2, 131–3.

Christensen, M. (2011) 'Discursively Shaping the Environment: Swedish National and Regional Media Coverage of the United Nations Climate Change Conference in Copenhagen ("COP15")', Paper presented to the Global Communication and Social Change Division of the International Communication Association (ICA) Conference, May 2011, Boston.

D'Angelo, P. (2002) 'News Framing as a Multi-Paradigmatic Research Program: A Response to Entman', *Journal of Communication*, 52:4, 870–88.

De Vreese, C. H., J. Peter and H. A. Semetko (2001) 'Framing Politics at the Launch of the Euro: A Cross-National Comparative Study of Frames in the News', *Political Communication*, 18:2, 107–22.

Djerf-Pierre, M. (2012) 'The Crowding-out Effect: Issue Dynamics and Attention to Environmental Issues in Television News Reporting over 30 Years', *Journalism Studies*, 13:4, 499–516.

Entman, R. M. (1991) 'Framing U.S. Coverage of International News: Contrasts in Narratives of the KAL and Iran Air Incidents', *Journal of Communication*, 41:4, 6–27.

—— (1993) 'Framing: Toward Clarification of a Fractured Paradigm', *Journal of Communication*, 43, 51–8.

—— (2004) *Projections of Power: Framing News, Public Opinion, and U.S. Foreign Policy* (Chicago: University of Chicago Press).

Eskjaer, M. (2009) 'Communicating Climate Change in Regional News Media', *International Journal of Climate Change Strategies and Management*, 1:4, 356–67.

Fraser, N. (2005) 'Transnationalizing the Public Sphere' (http://republicart.net/disc/publicum/fraser01_en.htm, date accessed 12 December 2008).

Gamson, W. A., and A. Modigliani (1987) 'The Changing Culture of Affirmative Action', in R. A. Braumgart (ed.), *Research in Political Sociology*, vol. 3 (Greenwich, CT: JAI) 137–77.

Gelbspan, R. (2004) *Boiling Point: How Politicians, Big Oil and Coal Journalists, and Activists are Fueling the Climate Crisis – and What We Can Do to Avert Disaster* (New York: Basic Books).

Gladwell, M. (1996) 'The Tipping Point', *The New Yorker*, 3 June (http://www.gladwell.com/1996/1996_06_03_a_tipping.htm).

—— (2000) *The Tipping Point: How Little Things Can Make a Big Difference* (New York: Little, Brown and Co.).

Goffman, E. (1974) *Frame Analysis: An Essay on the Organization of Experience* (New York: Harper & Row).

Habermas, J. (1989) *The Structural Transformation of the Public Sphere: An Inquiry into a Category of Bourgeois Society* (Cambridge, MA: MIT Press).

—— (2006) *Time of Transitions* (Cambridge: Polity).

Holland, M. M., C. M. Bitz and B. Trembley (2006) 'Future Abrupt Reductions in the Summer Arctic Sea Ice', *Geophysical Research Letters*, 33, L23503.

Hutchison, P. J. (2008) 'Journalism and the Perfect Heat Wave: Assessing the Reportage of North America's Worst Heat Wave', July–August 1936. *American Journalism*, 25:1, 31–54.

IPCC (2007) *Climate Change 2007: Synthesis Report*. Geneva: Intergovernmental Panel on Climate Change (IPCC).

Lyytimäk, J. and P. Tapio (2009) 'Climate Change as Reported in the Press of Finland: From Screaming Headlines to Penetrating Background Noise', *International Journal of Environmental Studies*, 66:6 (December), 723–35.

Nelkin, D. (1995) *Selling Science: How the Press Covers Science and Technology*, rev. ed. (New York: W. H. Freeman).

Nerlich, B., N. Koteyko and B. Brown (2010) 'Theory and Language of Climate Change Communication', *WIREs Climate Change*, 1, 97–110.

Nisbet, M. (2009) 'Communicating Climate Change: Why Frames Matter for Public Engagement', *Environment*, 51:2, 12–23.

Nuttall, M. (2012) 'Tipping Points and the Human World: Living with Change and Thinking about the Future', *Ambio*, 41:1, 96–105.

Olausson, U. (2009) 'Global Warming – Global Responsibility? Media Frames of Collective Action and Scientific Certainty', *Public Understanding of Science*, 18, 421–36.

Reese, S. D., and B. Buckalew (1995) 'The Militarism of Local Television: The Routine Framing of the Persian Gulf War', *Critical Studies in Mass Communication*, 12, 40–59.

Rick, U. K., M. T. Boykoff, and R. A. Pielke, Jr. (2011) 'Effective Media Reporting of Sea Level Rise Projections: 1989–2009'. *Environmental Research Letters* 6(1), accessed online on 5 February 2012.

Russill, C. and Nyssa, Z. (2009), 'The Tipping Point Trend in Climate Change Communication', *Global Environmental Change*, 19(3): 336–44.

Scheufele, D. A. (1999) 'Framing as a Theory of Media Effects', *Journal of Communication*, 49:4, 103–22.

—— (2000) 'Agenda-Setting, Priming, and Framing Revisited: Another Look at Cognitive Effects of Political Communication', *Mass Communication & Society*, 3, 297–316.

Semetko, H. A., and P. M. Valkenburg (2000) 'Framing European Politics: A Content Analysis of Press and Television News', *Journal of Communication*, 50:2, 93–109.

Trumbo, C. W., and J. Shanahan (2000) 'Social Research on Climate Change: Where We Have Been, Where We Are At, and Where We Might Go', *Public Understanding of Science*, 9:3, 199–204.

Weart, S. R. (2003) *The Discovery of Global Warming* (Cambridge, MA: Harvard University Press).

Wilkins, L., and P. Patterson (1991) *Risky Business: Communicating Issues of Science, Risk and Public Policy* (Westport, CT: Greenwood Press).

3
Eyes on the Ice: Satellite Remote Sensing and the Narratives of Visualized Data

Nina Wormbs

Introduction

'A picture is worth a thousand words' is a proverb in many languages. Behind it lies an understanding of how an image can at the same time capture and convey sentiments and complex information while speaking to our emotions and playing on our cultural knowledge of reading the visual. A picture is believed to have the ability of summarizing and explaining things in a straightforward way. The response of the viewer should ideally be 'I see'.

Satellites have been the eyes of the world for a number of decades. With distance to their object they are not only in the position of offering a synoptic and encompassing view of things impossible from the standpoint of humans (Parks, 2005), they also give an outsider's image, equipped with high-technology machinery and sensors, undisturbed by atmosphere and serenely placed in orbit according to the laws of physics. They inhabit space in a special way, and their position is exceptional. With the decline of human space flight, what else is there to watch the Earth from above?

This article deals with the satellite information systems that underlie our understanding and knowledge of the sea-ice minimum, and how that information has been displayed to the wider audience. The aim of the article is twofold: partly I want to discuss the satellite data and how it has been assembled over time and compared with other data, and partly I want to analyze some of the choices made when this data is visualized and subsequently used in news reporting, since a picture is worth more than a thousand words.

A strong illustration of the forcefulness of images is the famous Blue Marble photograph of the Earth from Apollo 17 in 1972 (Poole, 2008). That image, picture AS17-148-22727 in the NASA archive, was taken with a Hasselblad camera by one of the astronauts. Denis Cosgrove and Robert Poole have convincingly argued that it has been important not only for our understanding of the Earth as a planet, and thus for our self-understanding, but also that it was almost unique in doing that. The only other photograph taken by humankind from space and achieving the same status was the Earthrise picture from the Apollo 8 mission in December 1968, where the Earth appears to rise from behind the moon, which the mission had just circled. These two pictures illustrated that the Apollo program was not only about discovering and exploring the Moon, or perhaps not even primarily that, but instead a journey to discover and see the Earth (Cosgrove, 1994; Poole, 2008). The Blue Marble showed Earth as a vulnerable system in a hostile environment; it displayed it as a whole planet and not just patches of landscape. The fact that it was taken at a distance and from above rendered it 'objective' as so many images from above before it, however not as far away (Ekström, 2009).

My argument is that the images used in the news reporting on the sea-ice minimum carry with them connotations of place and importance that is of central interest to how we understand the sea-ice minimum as a global or a local issue, and subsequently how we understand climate change. These images help create narratives that are of a different kind from the ones we have associated with the Arctic historically (Sörlin and Bravo, 2002). And how these narratives inform our understanding is in turn connected to how we perceive our own role and our willingness to act in relation to climate change, and even though it is not evident precisely how visualizations in a broader sense affect this process, it is clear that media coverage is important in conveying messages of change (Nicholson-Cole, 2005).

Images illustrating scientific findings come in many forms and can sometimes be hard to understand, quite contrary to the intent of the producer of the image. Partly this is related to the fact that reading images is a cultural and historical practice that runs deeply (Berger, 1972). Images have been used to convey messages and in that process they have in turn been imbued with meanings which affect our reading. We associate and make references, often subconsciously, and the resulting understanding is more than could have been achieved with text alone. Moreover we read different images differently. Our understanding of the authenticity of the photograph goes back to its early use in the nineteenth century and the later acceptance of it as truth in court, resting on the assumption that the photographer was a 'witness' and the camera a mere technology (cf. Mnookin, 1998). Also in the realm of science, the removal of human agency from the production of the image was part of what rendered the photograph objectivity beyond the handmade illustrations (Daston and Galison, 1992).

Remote sensing images are often mosaics put together from a great number of different images, constructed from data assembled with various sensing technologies. This fact is almost always clearly declared when data centers such as the National Snow and Ice Data Center, or space agencies such as the European Space Agency, release images to the larger audience. However, even if the caption is correct and the journalist is transparent and clear with the data source, the image itself can still be read in different ways, and I want to explore the consequences of these different readings for the overall message conveyed. The fact that trust varies with different images complicates this reading even further.

Starting with a short history of remote sensing I move to the collection of sea-ice data by remote sensing, touching upon the other records of sea-ice distribution and thickness that are available. I then look at a few examples of different ways in which data extracted from the satellites were visualized, how those visualizations can be interpreted, and what the consequences might be.

Satellites as eyes on the world

How do we know that the sea-ice minimum of 2007 was a minimum? The short answer is that satellite sensors send data to computer systems, which in turn produce information on the state of the Earth. Since we have done this for some time now, we can detect change. As Paul Edwards has shown, the infrastructure of the information system for weather and climate data is an enormously complex 'machine' demanding standardization on almost every level of knowledge production (Edwards, 2010). Knowledge of the Arctic sea ice belongs in the same realm of activities, even though the Arctic is just one region of the planet. Arguably, however, and as discussed in Chapter 5, it is a region central to the global climate.

Data underlying the knowledge of the sea-ice minimum come primarily from satellites launched to study and observe the Earth. The science of remote sensing does not require satellites; it means only that there is a distance between the sensor and the sensed. Early examples of remote sensing were photographs taken from balloons, kites, or even by pigeons, but during most of the last century cameras were carried by airplanes (Campbell, 2002). Airplanes are still important since they are relatively easy to work with and can offer higher resolution due to their lower flying routes. With the advent of satellites came the possibility of covering larger and more remote areas of the Earth, and, perhaps more importantly, to get uninterrupted and very long-time series of data due to the fact that satellites can stay in orbit much longer than airplanes can fly (Cracknell and Hayes, 2007). In this chapter, the focus is on satellite remote sensing, as it has become one of the more important ways of knowing about Arctic sea ice, and will deal with some of the institutional context to the technology.

The satellite era began on 4 October 1957 when Sputnik was placed in orbit by the Soviet Union. The launch was a planned part of the ongoing International Geophysical Year of 1957–1958, an international scientific effort with historical roots, and with political causes as well as consequences (Needell, 2010). The so-called Sputnik shock spurred not only the US space program but also other high-technology and scientific efforts. By the end of the 1960s the Americans had won the space race and NASA had established a hegemonic role as a space agency (McDougall, 1985). The men on the moon were one sign of that, but during the same decade communications and weather satellites had also been launched and put into service, albeit with some difficulty (Mack, 1990, pp. 15–19). Not only were there technological thresholds to surmount, but perhaps more importantly the organization of space efforts was unclear. On the communications side private industry was heavily involved early on, which stirred conflict at the international level as space could be considered a common good and not something to commercialize.

The first American weather satellite, Tiros, was launched in 1960 and was soon followed by several generations. The Nimbus series, from the first one in 1964 to the seventh one launched in 1978 and in operation until 1994, was an important platform for experiments, but also produced valuable data. The first satellite solely devoted to remote sensing did not enter orbit until 1972. According to historian Pamela Mack, an important explanation for this delay of Landsat was the difficulty of involving the users of the data in the production of the system. This was not solved just because the satellite was launched and in orbit. In fact, Mack argues, not even in the late 1980s was the Landsat system a profitable endeavor with clearly defined users. NASA was good with research but not with application (Mack, 1990).

The Europeans gathered resources in the space business not only for economic but also for political reasons. A perceived technology and science gap with the United States could most easily be bridged by European cooperation and the space efforts were initially inspired by the way particle physics had been organized in the form of CERN in Geneva. After some turbulence the initial division of space efforts into a research organization and a launcher organization, which some of the neutral countries opted out of for political reasons, were abandoned and the present European Space Agency was reconstituted in 1972 (Krige et al., 2000a). John Krige has showed that the same difficulties pointed to by Mack and pertaining to profitability of applications' satellites were at hand also when ESA embarked on a satellite project for meteorology (Krige, 2000b).

ESA used the Landsat data and established a northern ground station to acquire data at the already existing space center Esrange above the Arctic Circle in northern Sweden. However, the organization itself did not at this point embark on a remote sensing project. Instead the first European remote sensing satellite was a French project with support from Sweden and Belgium. The first SPOT satellite, short for Satellite Pour l'Observation de la Terre, was launched in 1986. Hence in the late 1980s there were two systems in continuous use for Earth observation. Both of them launched several satellites to replace and complement the older ones and the systems are still in service. The first ESA remote sensing satellite, ERS, was launched in 1991.

The SPOT system also had difficulties with profitability. One of the partners, the Swedish Space Corporation, created a subsidy called Satellite Image to handle interpretation and commercialization of the data. Computer systems were in general expensive and a lot of power was needed to store and assemble the information, a problem equally true for the meteorological satellite systems put in place around the same time (Edwards, 2010). Moreover, interpretation of data was difficult and demanded highly skilled workers, which meant that images were not an off-the-shelf item. The Swedish Satellite Image struggled as a nonprofitable subsidy, despite its unique position having access to both Landsat and SPOT images, and from 1987 also images from the Japanese satellite MOS-1, and despite the fact that images produced were actually used (Wormbs and Källstrand, 2007).

In sum, remote sensing by satellite was well established at the end of the 1980s, but it had not yet become the profitable application that so many believed possible 20 years earlier. This development continued during the 1990s and well into the twenty-first century. However, launching of satellites continued, eased by cheaper technology and with continuous state support. These new satellites were essential for the monitoring of the 2007 sea-ice minimum, but for reference and in order to pinpoint change, older records were also of importance.

Collecting data on the ice at a distance

What data are the satellites collecting, and what are the challenges in that collection and the computation of information? There are a number of ways to gather data with sensors. In general the equipment can be divided into active and passive sensors. An active sensor transmits a specific signal, which is reflected in various ways depending on the surface and then registered by the sensor, much like normal radar. A passive sensor, on the other hand, registers the continuous emissions of radiation emanating from the Earth, such as heat or reflected sunlight, resembling a camera sensitive for radiation outside of the visible spectrum. Early satellites were often equipped with just one or two instruments, whereas more recent platforms can house a number of sensors catering to different needs (Rosengren, 2011).

The Envisat satellite, launched by the European Space Agency in 2002 and actively collecting data until 2012, for example, carried no fewer than 10 instruments, including those involved in guidance and control (ESA, 2012). Of central importance to monitoring the ice was the Advanced Synthetic Aperture Radar (ASAR) instrument. It operated in the so-called C-band, which is in the microwave area with frequencies of a few giga Hertz. Being a radar instrument it transmitted a signal which was reflected at the Earth and then picked up again by the satellite. The great advantage of radar for this purpose is that it permeates clouds and can be used both during night and day since it is not dependent on sunlight.

Of great importance to the collection of data is of course the satellite's orbital position, since that affects its usefulness. Weather and communication satellites are often placed in geosynchronous orbit around the equator since the interest there lies in continuously covering a certain area of the globe. Due to the laws of physics, it is, however, not possible to have a geosynchronous orbit in any plane other than that of the equator. Hence places not near the equator have to be observed from other orbits. Polar orbits are one good option as they allow for coverage of most of the globe, piece by piece, in continuous swathes. They normally have an orbital height of 700 to 900 kilometers, which is much lower than the geosynchronous at 36,000 kilometers and which of course affects the resolution of the data gathering. Polar orbits can be sun-synchronous and measure the reflected light from the sun (Rosengren, 2011; Cracknell and Hayes, 2007).

To be of any use, however, the data needs to be transmitted to Earth via radio. Ground stations are strategically placed to be able to receive the data, in order for the satellite to empty its memory and gather new data. A ground station placed closer to one of the poles has more chance of picking up the signals from a satellite in polar orbit since the satellite passes over the region of the station more often than it does over places closer to the equator. The actual download has historically been a limitation for the expansion of the technology. In the early days of remote sensing the downlink could perhaps manage a bandwidth in the megabyte range. Today that has increased a thousandfold and resulted in new possibilities for data handling. This handling has faced similar challenges over time since interpreting the information from the given sensored data is extremely computer-intensive (Rosengren, 2011). The collected information can be transformed into images that can be more easily interpreted. These images can in turn be put together to form mosaics where sequential images are made into swaths which are organized as larger images covering greater areas. In the case of the Arctic Ocean, mosaics are the normal way of getting an encompassing image due to the size of the area.

Satellites have monitored sea-ice extent continuously since the late 1970s. However, even the satellites launched in the 1960s collected records, although they could not be interpreted at the time due to lack of methods and computer power. These could be useful in order to expand the satellite records of sea ice, and ongoing projects are trying to decipher the data from three different Nimbus satellites in the 1960s to further the knowledge of the sea ice historically (NSIDC, 2010, Meier, Gallaher and Campbell 2013). Data was also gathered during the 1970s, but is not yet fully incorporated into the reference data.

Before satellite data was available submarines collected data on sea ice during their clandestine voyages in the Arctic Sea during the Cold War. The methods used differed, as did the temporal and spatial resolution, which posed challenges to analyzing the data. Moreover, the routes of the submarines did not coincide, with few exceptions, making comparisons over time difficult but not impossible. Several studies discussed the thickness of the sea ice based on unclassified data from upward sonar measurements (LeSchack et al., 1971; Swithinbank, 1972; McLaren, 1989). In 1998, however, data that was also classified was released and was updated in 2006 and subsequently available for research (NSIDC, 1998). The new data spurred activities and resulted in a number of articles claiming that the Arctic sea-ice cover was now thinner than in the late 1950s (see for example Rothrock, Percival and Wensnahan, 2008).

As discussed in Chapter 4, monitoring of sea ice has of course been of interest for centuries as it affects central economic and military interests not only for the Arctic peoples (Bravo, 2009). In trying to get a long record for comparison, a great number of disciplines are involved in making different data comparable (Polyak, 2010).

Displaying the sea-ice minimum of 2007

It is an interesting coincidence that both the International Polar Year of 2007–2008 and the International Geophysical Year of 1957–1958 could display the power of satellite technology not only to its users but also to a wider audience. During the most recent IPY satellites of different types and with different instruments collected information that showed the decreasing sea ice of the Arctic and the minimum of September 2007. IPY was a tremendous gathering of scientists and data and all the satellite space agencies around the world cooperated in monitoring the poles and making data available. When the sea ice in the Arctic reached its minimum, data and images of the event were abundant. Below I discuss three examples taken from three different news stories around the sea-ice minimum.

The first image I want to discuss is that featured in the *Telegraph* of 22 September 2007, titled 'Arctic sea ice "melts to all-time low"' (Figure 3.1).

The caption read 'Photographs taken in September 2005 and 2007' (Beckford, 2007). The imagined photographer had a distant view of a cloud-free globe in a dark universe where the orientation towards the sun is unclear. Still, the globe with its thin layer of atmosphere and bulging spherical properties led us to believe the caption that this might well be a photograph which we could have taken ourselves, had we been there. The caption did not give any source for these images, no 'photographer' or copyright. However, the images were the same as those used in a Reuters story 10 days later, and where the source was impeccably accounted for: 'A combination photo of NASA satellite images from September 21, 2005 (top) and September 16, 2007 (bottom) and released on September 21, 2007 shows Arctic summer sea ice coverage in 2005 and 2007 respectively' (Dunham, 2007). Moreover, the lower picture had its own caption in the slide show offered by Reuters where the details of the image were further displayed. The image was a NASA satellite image from 16 September, showing Arctic summer sea ice and citing scientists attributing the decline to human-caused warming. In the Reuters reproductions the images were not claimed to be photographs but were

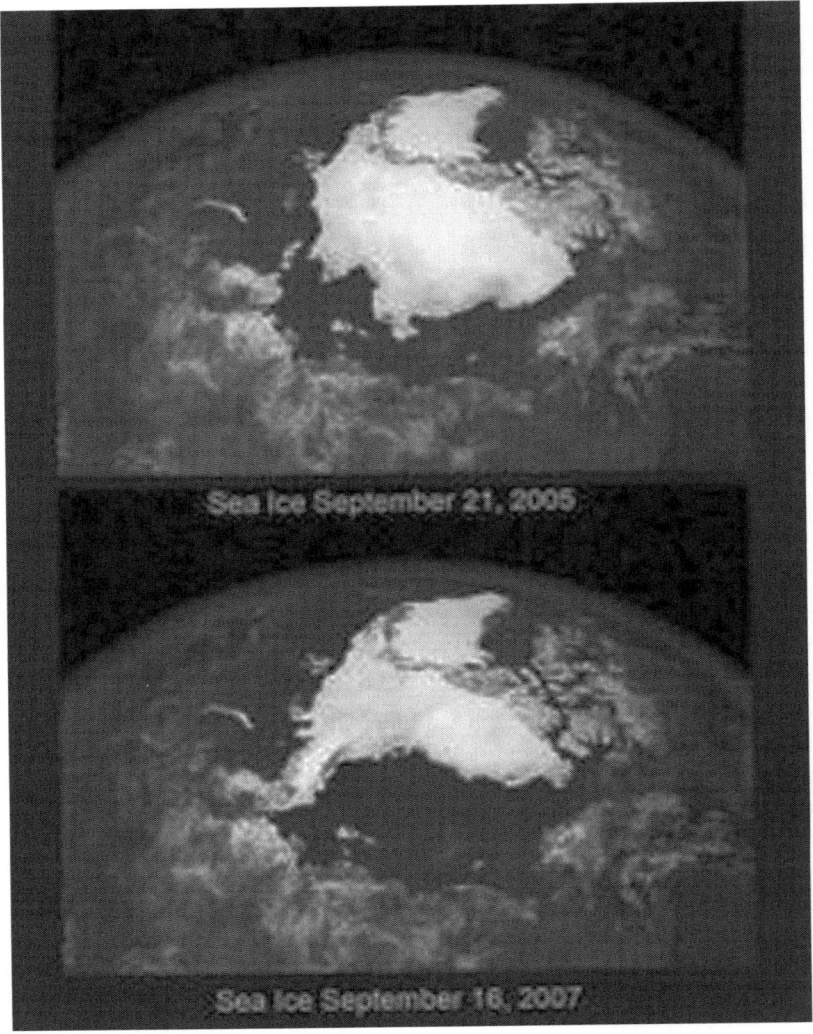

Figure 3.1 Arctic sea ice 'melts to all-time low' (2005 and 2007)
Source: NASA.

instead clearly stated to be images. The translation by the *Telegraph* journalist is obvious and striking. In removing some of the information, the meaning of the illustration was altered and the way in which readers could understand the information was changed too.

The *New York Times* made another choice in its article with the title 'Arctic Melt Unnerves the Experts', published 2 October 2007 (Revkin, 2007). The article was topped with a photograph taken from an icebreaker showing open waters and ice in the Arctic Ocean. The retreating sea ice was illustrated in a multimedia interactive graphic, which naturally worked only on the Web. The graphic showed a polar map with the Bering Sea in the north and the Shetland Islands in the south, and thereby the expected orientation from a European perspective. The graphic pointed out the Northern Sea Route as well as the Northwest Passage and indicated the average summer minima from 1979 to 2005. The ice extent could be seen to change for the summers between 2003 and 2007. The map, a clear construction of humankind with deep historical roots and often known to display an idea of things or a claim rather than being a true representation (Harley, 2001), made all those clarifications possible and resulted in an educating effort in which readers could spend a number of minutes exploring and understanding change. Interestingly enough, however, data sources were not revealed in the graphic.

Yet another take on the sea-ice minimum was chosen by *National Geographic News* online on 17 September 2007. In an article titled 'Arctic Melt Opens Northwest Passage' the consequences of open waters and new sea routes were discussed and the geopolitical implications were problematized (Roach, 2007). The minimum itself had been dealt with as a prognosis already in August and this one repeated the news but also focused on the passage. The chosen illustration was a mosaic from the European Space Agency, ESA (Figure 3.2). The image was clearly stated to be a mosaic and the black hole above the North Pole, due to the satellite orbits, was not filled but left black in accordance with data. Mosaics can be done differently, but in this one the satellite tracks were clearly visible and it was obvious that the image was put together by the straps of data that had been assembled at each passing over the area. In leaving these straps visible, the reader could see that the image was composed of several images and thereby the distance to the data was reduced. The text also revealed that the mosaic was made out of 200 images and disclosed to the reader some of the work such a composite might entail and thereby further underlined the scientific basis for the image.

The image itself was taken from ESA where the same image also illustrated the central piece on the sea-ice minimum (ESA, 2007). The title was 'Satellites witness lowest Arctic ice coverage in history', alluding to the position of the satellite as eyes on the globe, ready to record and serve as providers of evidence of the event. In the ESA piece more information

62 Nina Wormbs

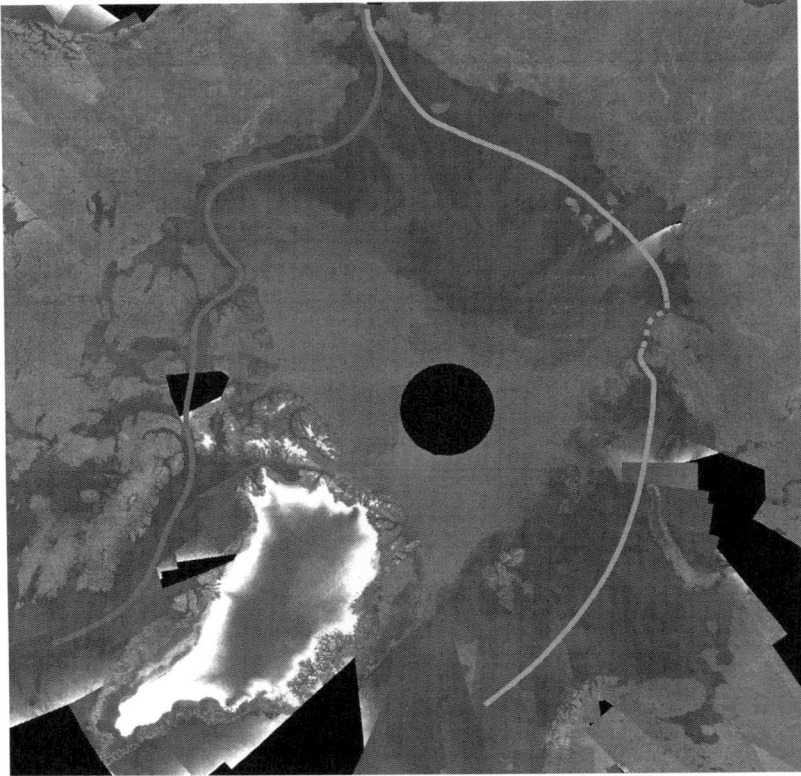

Figure 3.2 Satellites witness lowest Arctic ice coverage in history
Source: ESA.

was given about the satellite and its instrument as well as how ESA was taking part in the IPY efforts and what launches lay ahead.

The *Telegraph*, *New York Times* and *National Geographic News* all chose different ways of illustrating the sea-ice minimum based on the information they had received from scientists working in data centers or space agencies. I argue that these different choices also affected how those illustrations were read and consequently what larger message was conveyed.

As hinted above, the ESA picture is the one closest to data, in the sense that some of its pieces are visible to our eyes. However, the original recorded data is still not what we see. Instead the stripes are computational visualizations of enormous amounts of data collected by the sensors and run through computers. The image is the way in which the scientists, and the public, can make sense of the information and is therefore not just an illustration of a physical state, but also the tool to understand that process. Remote sensing images are hence not mere popularizations and simplifications of science, which we see so often and which can undergo critique for being just that (Hilgartner, 1990).

The polar map used by the *New York Times* merges an old visualization technology with new scientific findings. The witness in this image is the reader, and the data behind the displayed information is actually not disclosed, even though a number of scientists figure in the text. The message is straightforward enough. At the same time, a polar map of the Arctic also carries cultural meaning which is overlaid on the image of the shrinking ice. Polar maps are surprising for those who have not seen them, as the poles are not normally in the center but stretched out on the periphery. But they are also common in Cold War history, illustrating the distance between the Soviet and the United States, depicting a militarized zone with distant early warning systems and places for launch pads. In this map of the shrinking sea ice, the references to those geopolitical processes are the sea routes that are made possible by climatic change.

The *Telegraph* image, also used by Reuters but with a clearer caption, is further away from the data than the ESA image, without being a popularization. The shrinking sea ice is placed on a sphere, itself constructed by satellite data, and thereby it resembles what we might be able to see were we placed in space and able to look for ourselves. Moreover, the two images construct a sequence and we can understand how the next image of sea ice might look in a few years, an efficient way of conveying a message (Kjeldsen, 2011).

The *Telegraph* image refers to the Blue Marble, but does not cite it fully. To begin with, there was no sunshine orienting the globe in the solar system, confusing those who read images of the globe against a black background as the Earthrise images or the original Blue Marble, but perhaps confirming for others that this was indeed another type of image. Second, there are no clouds which we have learned to associate with Blue Marbles. We know that the atmospheric layer will never be perfectly transparent over such a large area, and thus their absence might in the same way either puzzle us or serve as a confirmation of the type of image we are looking at. However, the association drew the viewer into an understanding of the Arctic and its shrinking sea ice as part of the whole Earth. Contrary to the straps in the ESA image, or the geopolitical polar map, the marble image positioned the Arctic on the actual globe in relation to the rest of the sphere, hence the message that the shrinking Arctic sea ice was also important for the entire globe was amplified.

However, the association to the photograph is so strong that people might think that the image is a photograph and judge its credibility from that vantage point. One or another journalist might make the same mistake, as we have seen, and contribute to this misconception. Transparency and accountability of science are central for the credibility of science-based policy. Given the polarized climate debate in certain parts of the world, however, this might be truer in this area than in most. A misunderstanding of the nature of a certain image might, when revealed as such, lead to a blow to public confidence in the underlying scientific knowledge production. Paradoxically, the image that can best convey the message of the importance of the shrinking sea ice could also be the one running the highest risk of being misinterpreted and in the worst case undermining the entire scientific message.

Blue Marbles and the blurring of data sources

In early 2012, NASA released what they called two new Blue Marbles, one of the Eastern Hemisphere and one of the Western (NASA, 2012). The reference to the original was clear, particularly as the angle from which the Eastern Hemisphere was portrayed is similar to that from which the original 22727 was taken. These 'marbles' were put together with data from one specific satellite, the Suomi NPP, in a polar orbit since 2011, at one specific flyover in January 2012. Information on how the 'high-resolution' image was constructed was easily found on the NASA web page, but at the same time the tribute to the original photograph was underscored in the name and in the text, resulting in a tension between messages conveyed.

These two are not the first follow-up Blue Marbles that NASA has constructed. In 2002 the space agency released a 'marble' building on images from several satellites. At the center is the west coast of the North American continent, and covering it are beautiful changing clouds that swirl and seem to move across continents and oceans, as we expect them to do. The visualization experts working at NASA, first and foremost Reto Stöckli and Robert Simmon, are credited with the work, which was done in stages (NASA, 2011). Innumerable data sets were put together by Stöckli to form a rectangular world map at high resolution. To this map Simmon added extra color in the oceans (based on satellite data on chlorophyll) and snow and ice at the poles. It was then turned into a

Figure 3.3 Left, original AS17-148-22727 from Apollo 17 in 1972. Right, NASA Blue Marble from 2002.

Source: NASA. http://earthobservatory.nasa.gov/blogs/elegantfigures/2011/10/06/crafting-the-blue-marble/

globe using a software program, after which a few layers of clouds with different transparency and highlights were added and subsequently tweaked to get the almost-final image. The last touch involved some Photoshop work, where a few clouds were repeated over the Pacific Ocean (NASA, 2011).

This so-called Photoshopping is something a trained eye can easily spot (and eyes have) and immediately the 'photograph' is unveiled as a mere image or even the artist's impression. At the same time it is clear from the NASA web page that most of the information in the image is based on satellite sensing data meticulously incorporated into a frame that we can recognize as the Earth. Simmon also explains that the reason for using Photoshop was missing data at the equator due to the orbits of the satellites. These gaps had to be filled to avoid white streaks, and the mistakes are listed at the end of the account given by Simmon on the NASA web page (NASA, 2011). As far as the ice goes, he further stated, 'I also had to add a stand-in for sea ice, since it's impossible to measure chlorophyll beneath a few meters of snow and ice. At least that was simple – I just replaced missing data near the poles with white. In addition to the sea ice, I brightened and reduced the saturation of Antarctica, which was pasted into the original from a different dataset' (ibid.).

This Blue Marble was the one lighting up every iPhone when first switched on and is still a widely used marble. And the image illustrates the blurring of the data sources that has become commonplace. In the case of the Simmon Photoshopping the alternations were harmless as the image he constructed might well have been possible. However, if remote sensing is to be Earth observation, we need to be certain that what the instruments have 'seen' is in fact what we have been shown.

The knowledge of the sea-ice minimum is built primarily on remote sensing data, which could be compared over several decades. Moreover, the summer of 2007 was not the first time that the ice was at a low during the preceding decade. It dropped in 2003 and then again in 2005 before the extreme of 2007 was confirmed. To understand that the ice was shrinking was a process taking place over time and involving a great many people, just as was the process of imbuing the original Blue Marble with the meaning of the Whole Earth, as a vulnerable and precious unity in need of care.

Conclusion

The aim of this chapter has been to show the complexity of gathering remote sensing data of Arctic sea ice and then point to some of the narratives that these data can be placed into with the aid of images. I have argued that the choice of illustration is of importance for which narrative is implied in the overall news story. Since we as readers tend to search for cause and effect in our efforts to create meaning, possible consequences of certain changes are implied in the chosen illustrations, and in turn they support specific stories. However, the way in which, for example, remote sensing imagery is presented can also have a crucial importance for how the reader appreciates the underlying scientific data. Easy-to-make mistakes as to the source of data and the production of the image can undermine the overall scientific conclusion.

Cosgrove has pointed to the forcefulness of the orientation of the original Blue Marble, AS17-148-22727. The 'dark continent' of Africa was in the center of the Earth, clearly visible and bathing in sunlight, the recently discovered continent of Antarctica at the bottom firmly frozen as it were, and the old and the new worlds barely appearing as a narrow band on the periphery, challenging prevalent perceptions based on Western cartography. Similarly, a polar map is an eye-opener at first sight because it changes the perspective and makes one realize what should be obvious: just as there is no given orientation for a polar map, there is no center on the surface of a sphere. In the same way as 22727 drew Africa and Antarctica into the whole Earth, the image of the melting Arctic displayed on a computer-generated blue marble might position the region more firmly in our consciousness not only as a stretched-out periphery, but just as easily a center.

References

Beckford, M. (2007) Beckford, M. 'Arctic sea ice "melts to all-time low"', *The Telegraph*, 22 September (http://www.telegraph.co.uk/earth/earthnews/3307816/Arctic-sea-ice-melts-to-all-time-low.html, date accessed 17 February 2013).
Berger, J. (1972) *Ways of Seeing* (London: BBC).
Bravo, M. T. (2009) 'Sea Ice Mapping: Ontology, Mechanics and Human Rights at the Ice Floe Edge', in D. Cosgrove and V. della Dora (eds), *High Places: Cultural Geographies of Mountains, Ice and Science* (London: I. B. Tauris).
Campbell, J. B. (2002) *Introduction to Remote Sensing* (London: Taylor & Francis).
Cosgrove, D. (1994) 'Contested Global Visions: One-World, Whole-Earth, and the Apollo Space Photographs', *Annals of the Association of American Geographers*, 84:2, 270–94.
Cracknell, A., and L. Hayes (2007) *Introduction to Remote Sensing* (London: CRC Press).
Daston, L., and P. Galison (1992) 'The Image of Objectivity', *Representations*, 40 (Autumn), 81–128.
Dunham, W. (2007) 'Scientists See Dramatic Drop in Arctic Sea Ice', Reuters, 22 October (http://uk.reuters.com/article/2007/10/02/uk-arctic-ice-idUKN0134058320071002, date accessed 17 February 2013).
Edwards, P. (2010) *A Vast Machine: Computer Models, Climate Data and the Politics of Global Warming* (Cambridge, MA: MIT Press).
Ekström, A. (2009) 'Seeing from Above: A Particular History of the General Observer', *Nineteenth-Century Contexts*, 31:3 (September), 185–207.
ESA (2007) 'Satellites Witness Lowest Arctic Ice Coverage in History', 14 September (http://www.esa.int/esaEO/SEMYTC13J6F_index_0.html).
—— (2012) 'ESA Declares End of Mission for Envisat', 9 May (www.esa.int/esaCP/SEM1SXSWT1H_index_0.html, date accessed 3 November 2012).
Harley, J. B. (2001) 'Deconstructing the Map', in J. B. Harley, *The New Nature of Maps: Essays in the History of Cartography* (Baltimore: The Johns Hopkins University Press).
Hilgartner, S. (1990) 'The Dominant View of Popularization: Conceptual Problems, Political Uses', *Social Studies of Science*, 20:3, 519–39.
Kjeldsen, J. E. (2011) 'Visual Argumentation in an Al Gore Keynote Presentation on Climate Change' in F. Zenker (ed.), *Argumentation: Cognition and Community*. Proceedings of the 9th International Conference of the Ontario Society for the Study of Argumentation (OSSA), May 18–21, 2011. Windsor.
Krige, J. (2000b) 'Crossing the Interface from R&D to Operational Use: The Case of the European Meteorological Satellite', *Technology and Culture*, 41, 27–50.
Krige, J. et al., (2000a) *A History of the European Space Agency*, 2 vols. (Nordwiijk: ESA).
LeSchack, L.A., et al. (1971) 'Automatic Processing of Arctic Pack Ice Data Obtained by Means of Submarine Sonár and Other Remote Sensing Techniques', in

J. B. Lomax (ed.), *Propagation Limitations in Remote Sensing*, AGARD-CP-90–71, pp. 5-1–5-14, North Atlantic Treaty Organisation, Brussels.

Mack, P. (1990) *Viewing the Earth: The Social Construction of the Landsat Satellite System* (Cambridge, MA: MIT Press).

McDougall, W.A. (1985) *The Heavens and the Earth: A Political History of the Space Age* (New York: Basic Books).

McLaren, A. S. (1989) 'The Under-Ice Thickness Distribution of the Arctic Basin as Recorded in 1958 and 1970', *Journal of Geophysical Research*, 94:C4, 4971–83.

W. N. Meier, D. Gallaher, and G. G. Campbell (2013) 'New Estimates of Arctic and Antarctic Sea Ice Extent During September 1964 from Recovered Nimbus I Satellite Imagery', *The Cryosphere*, 7, 699–705.

Mnookin, J. (1998) 'The Image of Truth: Photographic Evidence and the Power of Analogy', *Yale Journal of Law and the Humanities*, 10:1 (Winter), 1–74.

NASA (2011) 'Crafting the Blue Marble', by Robert Simmon (http://earthobservatory.nasa.gov/blogs/elegantfigures/2011/10/06/crafting-the-blue-marble/).

—— (2012) http://www.nasa.gov/topics/earth/features/viirs-globe-east.html, date accessed 17 February 2013.

Needell, A. A. (2010) 'Lloyd Berkner and the International Geophysical Year Proposal in Context: With Some Comments on the Implications for the Comité Spéciale de l´Année Géophysique Internationale, CSAGI, Request for Launching Earth Orbiting Satellites', in R. D. Launius, J. R. Fleming and D. H. DeVorkin (eds), *Globalizing Polar Science: Reconsidering the International Polar and Geophysical Years* (New York: Palgrave Macmillan).

Nicholson-Cole, S. A. (2005) 'Representing Climate Change Futures: A Critique on the Use of Images for Visual Communication', *Computers, Environment and Urban Systems*, 29, 255–73.

NSIDC (National Snow and Ice Data Center) (1998, updated 2006) *Submarine Upward Looking Sonar Ice Draft Profile Data and Statistics*. Boulder, CO.

—— (2010) http://nsidc.org/monthlyhighlights/2010/02/techno-archaeology-rescues-climate-data-from-early-satellites/.

Parks, L. (2005) *Cultures in Orbit: Satellites and the Televisual* (Durham, NC: Duke University Press).

Polyak, L., et al. (2010) 'History of Sea Ice in the Arctic', *Quaternary Science Reviews*, 29, 1757–78.

Poole, R. (2008) *Earthrise: How Man First Saw the Earth* (New Haven: Yale University Press).

Revkin, A. C. (2007) 'Arctic Melt Unnerves the Experts', *New York Times*, 2 October (http://www.nytimes.com/2007/10/02/science/earth/02arct.html, date accessed 17 February 2013).

Roach, J. (2007) 'Arctic Melt Opens Northwest Passage', *National Geographic News*, 17 September 2007 (http://news.nationalgeographic.com/news/2007/09/070917-northwest-passage.html, date accessed 17 February 2013).

Rosengren, M. (2011) Interview with author 13 December 2011.

Rothrock, D. A., D. B. Percival and M. Wensnahan (2008) 'The Decline in Arctic Sea-Ice Thickness: Separating the Spatial, Annual, and Interannual Variability in a Quarter Century of Submarine Data', *Journal of Geophysical Research*, 113:C05003.

Sörlin, S., and M. Bravo (eds) (2002) *Narrating the Arctic: A Cultural History of Nordic Scientific Practices* (Canton, MA: Science History Publications).

Swithinbank, C. (1972) 'Arctic Pack Ice from Below', in T. Karlsson (ed.), *Sea Ice: Proceedings of an International Conference, Reykjavik, Iceland*, National Research Council, Reykjavik, cited in A. S. Thorndike, D. A. Rothrock, G. A. Maykut and R. Colony (1975), 'The Thickness Distribution of Sea Ice', *Journal of Geophysical Research*, 80:33 (November 20), 4501–13.

Widmalm, S. (1990) *Mellan kartan och verkligheten: Geodesi och kartläggning, 1695–1860* (Uppsala: Uppsala University).

Wormbs, N., and G. Källstrand (2007) 'A Short History of Swedish Space Activities' (Nordwiijk: ESA).

4
An Ice Free Arctic Sea? The Science of Sea Ice and Its Interests

Sverker Sörlin and Julia Lajus

The 2007 sea-ice minimum was quickly framed as a unique event and a very clear signal that the Arctic was a bellwether for global climate change. It became an event of the future it heralded rather than of the past or recent changes that had created it. That it was a single-year event was considered of minor importance; the 2008 to 2012 figures have corroborated the long-term downward trend and the US National Snow and Ice Detection Center that monitors the ice maintains that they expect a seasonally ice-free Arctic in 20 to 30 years.[1] The minimum of 2007 can be seen as an event of the Anthropocene: the kind of event which signals that humanity has now become, according to some scientists (Crutzen and Stoermer, 2000; Steffen, Crutzen and McNeill, 2007; Robin and Steffen, 2007) the most significant agent of change on the planet. In fact it is an event that is more significant than most others, which are punctual episodes with little or no lasting effects and with unclear longevity. Storms, hurricanes, floods, fires happen, cause great havoc and breaking news, but then they fall into oblivion. Melting sea ice is, perhaps paradoxically, more lasting. It has the effect that once it has melted it may be gone for a long time and a major shift back to a colder climate over a period of many years is required to restore it. Ice may form in an instant, a moment of crystallization that can even be heard as clearly as a shotgun sometimes, but its life cycle is full of inbuilt slowness. First-year ice is thinner and goes away much more easily than multi-year ice. Vanishing ice is, potentially, a multi-year event, likely even a multi-decadal event, for what it heralds. Perhaps its most salient impact is that it reminds us of the fundamentally human forcing of the environment, further pushing humanity into the era of 'the environmental' even at the poles.[2]

In this sense our contemporary sequence of sea-ice minima can be read as a 'cryo-historical' moment – *cryo* signifying ice and snow, directing our attention to the historical powers of human forcing in the Anthropocene – a moment that demonstrates humanity's hegemony over Earth as manifested in a retreat of one of its elements.[3]

In this chapter we shall use a distancing technique to approach the sea-ice minimum: time and temporality, and the attempts through science to use the evidence of ice in time to make political claims on, or pursue political acts for, the future of the Arctic. The master narrative is now that humanity (or rather its energy-intensive share of several billion people) has caused a melting trend of Arctic sea ice and thus broken a long Holocene stability (Rockström et al., 2009) with an Anthropocene disruption. If that is true, it would seem a first-order duty to examine the records of the Holocene and see what they can tell us in order to better understand our concerns with the vanishing ice as evidence of a current mega-transition. Was the 2007 minimum unique? When and why did science start to study Arctic sea ice? Have there been periods of an ice-free Arctic Sea in the past? And, if they did occur, how does it impact on interpretations of our present-time discourse on the possible emergence of an ice-free Arctic Sea, if at all?

Ice as an indicator of changes in climate is an old phenomenon, and retreating and expanding glaciers have been known since at least medieval times although people then had very little chance of bringing any meaningful pattern to the observations that they recorded or the lore of previous change. Glaciers started to be used as scientific climate indicators in the nineteenth century and a broad and institutionally organized cryospheric research took shape only in the twentieth century (Bader, 1949; Clarke, 1987). Climate change may, in retrospect, have appeared an obvious companion idea, but this relationship between ice and climate was rarely put forward as a serious alternative for the immediate future on the human timescale of decades, generations, or even centuries. But when it finally was, comparatively late in the middle of the twentieth century, sea ice was part of the story.

We start by visiting the idea of an ice-free Arctic in the past, then moving on to the scientific undertakings on sea ice in the Soviet Union, second to none. Interwar efforts outside the Soviet Union were only matched by Nordic researchers, with whom we deal subsequently. Finally we discuss the Cold War efforts and their military connections. That science is interest-driven is evident throughout the entire period. Sea-ice minima may comprise straightforward facts, but the underlying knowledge is the outcome of a complex science politics of circumpolar ice.

Long-term variabilities

In the 1920s, just a few years into the period of 'Early Arctic Warming' (EAW) in the European Arctic between 1919 and the 1940s,[4] climate scientists saw reason to believe that the Arctic Ocean had been free of ice for long time periods in the past. The issue had already been debated in geographical journals such as *Petermann's Geographische Mitteilungen* in the latter half of the nineteenth century and the US Navy hydrographer Silas Bent presented the idea that enormous amounts of heat was transported into the Arctic from southern latitudes leading to occasional thawing and a "supposed open Polar Sea" as illustrated in one of his publications (Bent, 1872). Swedish oceanographer Otto Pettersson had proposed in the early years of the twentieth century that there were long-period variations in the circulation of the oceans caused by changes in 'tide-generating force'. This force in turn varied with the declination and proximity of the sun and moon to the Earth and reached maxima every 1,700 years, roughly 3500 BC, 1900 BC, 250 BC and AD 1433 when tidal forces brought more warm water into the Arctic (Pettersson, 1905, 1914, 1915). How far these periodical warmings through currents of the Arctic Ocean would affect ice conditions was not obvious and was a matter of debate among scientists. However, the position that the Arctic had gone through periods of very little or no ice was common for several decades, reinforced by the ongoing warming.

Similar ideas lived on throughout this warming period, which saw a retreat of sea-ice in the late 1930s on the Atlantic and Russian part of the Arctic of the same magnitude as in the early 2010s (Overland et al, 2011). As late as 1949, when the period had just ended, British meteorologist and climatologist C. E. P. Brooks, citing Pettersson's 1914 paper, was open to the thought that the Arctic Sea was free of ice for very long periods of time, hundreds of years at a time: 'The final stage came about 500 BC when for some reason the Arctic ice-cap last became firmly established, apparently very extensively, after a few centuries of heat and drought' (Brooks, 1949, p. 327). As causes, he suggested sun spots, volcanic activity, and tidal forces. Brooks also had a quite advanced theoretical understanding of the temperature relations that were required to reach a 'glacial' or a 'non-glacial' regime and the differences of temperatures required were small indeed (Brooks, 1925). There was also solid evidence, in Brooks's view, that there had been periods in the geological past with significantly higher temperatures in the Arctic. Based on data by Austrian climatologist Fritz Kerner he proposed extremely high January temperatures at the 75th parallel on the level of present day levels in southern England: 'these temperatures are quite consistent with the remains of vegetation discovered by geologists' (Kerner, 1910; Brooks, 1949, p. 141).

The evidence that was accumulating to suggest a variable pattern between warm and cold periods never provoked any mention of human climate forcing or the possibility that humans would have anything to do with the causes of sea-ice changes, either in the past or in the future. The contrary was, however, present, that is that climate change affected humans, the 'environmentalism' trope which by the mid-twentieth century already had a considerable and not too venerable history. In the 1950s there was for example a climate-related suggestion proposed for the demise of the Norse settlement of Greenland (Utterström, 1955; Grove, 1988) during what was then, since 1939, called 'the Little Ice Age' (Matthes, 1939; Clague and Slaymaker, 2000; Fagan, 2000). Concern for contemporary populations linked to the reduction of sea ice did not occur until the final years of the twentieth century, when climate change became acknowledged on a broad basis.[5] This should come as no surprise. In the Western perspective sea ice has for hundreds of years, and not just since the sinking of the *Titanic* in 1912 (a year of unusual southern spread of sea ice in the Newfoundland area; Smith, 1932, p. 404), been regarded as a risk and an obstacle to the economic values of the Arctic, which lay in whaling, sealing, shipping, and mineral and fossil-fuel extraction.[6] This is the opposite of the value ice represents to Arctic populations like the Inuit, for whom the sea ice – hunting ground, livelihood, medium of travel, play and wayfaring – is awaited with expectation, and where November, the usual first month of sea ice, is called *Tusaqtuut*, 'the news season'.

An ice-free Arctic Sea and the northern expansion of Soviet power

The interest in melting and strangely behaving sea ice, along with commercial and geopolitical motives related to the Svalbard Treaty in the 1920s, resulted in a significant growth of publications in several countries, not least on research methods and terminological issues. Some of these publications mentioned the trope of the historically ice-free Arctic Sea.[7]

The possible waning of ice in the Arctic affected all circumpolar nations, but the news was most forcefully felt in the Soviet Union. There the drastic warming trend visible in the European Arctic in the 1920s prompted the new Bolshevik government to major and bold new policy directions, including the establishment of the agency for the Northern Sea Route, *Glavsevmorput*, in 1932.[8] The idea was straightforward: in a warmer climate the shipping period north of Russia would be longer, perhaps long enough to allow for substantial commercial shipping.

This major national interest, which was geared to secure trans-Arctic trade and shipping and serve as an outlet for the massive natural resources in the Soviet North, led to a substantial prioritizing of funds for Arctic research which put Soviet scientists far ahead of other nations in studies of sea ice. However, the Soviet scientists also upheld important connections with the Scandinavian countries, which complemented the Soviet knowledge in interesting ways. The Soviet Union has the longest coast in the Arctic Ocean, covering 180 degrees longitude, and sea ice was a major obstacle for establishing navigation. Nikolay Zubov, one of the leading Soviet sea-ice scientists of the entire period from 1920 through to the late 1940s, stressed that not only meteorological but oceanographical data were needed to forecast sea-ice distribution (Zubov, 1939, p. 5).

Understanding of oceanographic processes in the Arctic was a goal to which Zubov devoted his life. He was especially intrigued by the rapid warming of the seawater temperature. Zubov was one of the organizers of the Floating Maritime Research Institute, which in 1929 took over the regular oceanographic sampling along the Kola meridian section in the Barents Sea, where hydrographical data were collected from 1898 according to the program of the International Council for the Exploration of the Sea (ICES). Nikolai Knipowitsch, the leader of this research in the early period, already in 1921 suggested a warming of the Barents Sea when he explained his recent oceanographic data on the Kola meridian section, comparing them with data from the beginning of the century (Knipowitsch, 1921).

In the same year, another leading Soviet scientist, meteorologist and oceanographer, Vladimir Wiese, charted the border of distribution of sea ice in the western part of the Kara Sea from the deck of the vessel *Taimyr*. His main scientific efforts were directed towards understanding atmosphere–oceanic interactions, changes which could be revealed by observation of sea-ice distribution (Wiese, 1924).[9] In 1923, Wiese published an extensive paper, 'On the possibility of forecasting the state of ice in the Barents Sea', where he could show clear connections between inflow of warm Atlantic waters into the Barents Sea and ice conditions (Wiese, 1923). Inflow of these waters was affected by the atmospheric circulation, especially by the position of the Icelandic air pressure minimum. Using statistical methods Wiese gave a formula of dependence of ice distribution from the position of this minimum. Theoretically he built his work on the analysis provided earlier by the Norwegian oceanographer and explorer Fridtjof Nansen and his colleague Björn Helland-Hansen (Nansen and Helland-Hansen, 1909). Despite the fact that it was actually based on rather scarce data and thus provided only approximate calculations, Wiese through this study laid the foundation of Soviet sea-ice research for many years (Treshnikov, 1978). Further analysis of the phenomena related to 'the warming of the Arctic' provided by Zubov was based on direct observations of water temperatures and sea-ice distribution complemented by materials published by the Danish Meteorological Institute.[10] Zubov too referred to the path-breaking work by Nansen and Helland-Hansen: 'As early as 1909 Nansen and Helland-Hansen in *The Norwegian Sea* [Nansen and Helland-Hansen, 1909] had pointed out a correlation between the temperature of the North Cape Current off the Murman coast and the state of ice in the Barents Sea in the same year' (Zubov, 1933). Influence from Björn Helland-Hansen was crucially important for the training of Zubov. In 1913 he was a student of Helland-Hansen's when he attended the International Oceanographic Courses held in Bergen every summer since 1903. There Zubov received new ideas on dynamic oceanography and enthusiastically adapted and developed them. Soviet oceanographers very carefully followed publications on changing conditions in the polar regions and especially those that provided new methodological opportunities to understand the change.

Soviet scientists not only analyzed the patterns of diminishing sea ice, they made very reliable predictions. Thus, in the fall of 1929 and the spring of 1930 water temperatures along the Kola meridian were higher than usual and Zubov, who led these observations, predicted a smaller amount of ice in the Barents Sea for the summer and early fall of 1930. Based on these predictions he planned and fulfilled a successful expedition on a small Norwegian-built wooden ship, *Nikolai Knipowitsch*, that went up north in the strait between Spitsbergen and Franz Josef Land to the 81st parallel in almost total absence of sea ice. Zubov, not lacking a good sense of humor, sent a telegram to his institute: 'The ice edge is lost. Those who will find it, please, deliver it to the address: longitude 81, *Knipowitsch*' (Zubov, 1932, p. 30).

One of the reasons for Soviet scientists to propose inclusion of oceanography into the program of the Second IPY was an interest in tracing changes in the temperature in the Gulf Stream, which heavily affected the ice regime in the polar seas. This was officially done at the first meeting of the International Commission for the Preparation of the

Second IPY, which took place in August 1930 in Leningrad (Lüdecke and Lajus, 2010, p. 163). The most internationally known oceanographer at this meeting was Norwegian Harald Ulrik Sverdrup, who already had established connections and exchange of publications with Soviet scientists. One of them was Wiese, with whom he discussed a wide range of questions on polar oceanography, including sea ice, since at least the fall of 1927.[11] Soviet oceanographers hoped that international cooperation during IPY would facilitate sampling of data in both the Barents and Norwegian seas. However, due to the economic crisis other countries did not put additional finances into fulfillment of this proposal.

During the unusually warm years of the Second IPY, 1932 to 1933, when the ice cover of the Arctic Ocean diminished, Soviet scientists consciously used these phenomena to facilitate navigating into the high Arctic. The first-ever circumnavigation of Franz Josef Land, accompanied by a series of oceanographic samplings, was led by Zubov in 1932 (Zubov, 1933). Among other data on the 'warming of the Arctic' he noted the results of a Swedish–Norwegian expedition conducted in 1931 and led by Swedish glaciologist Hans Wilhelmsson Ahlmann.[12] Another expedition, which also used the favorable ice conditions of 1933, went along the northern coasts of Severnaya Zemlya islands. Finally the conquering of the passage along the coast of Siberia in one season was fulfilled. Scientific leader of the latter expedition on board the ice-ship *Sibiriakov* was Vladimir Wiese, at that time deputy director of the Arctic Research Institute founded in 1930 in Leningrad, who organized oceanographic and meteorological samplings along the route. Later, in 1934, he predicted diminishing ice for the expedition along the Northern Sea route in the opposite direction – from the east to the west – on board the icebreaker *Litke*, an expedition which he also led. Pre-World War II Soviet Arctic expeditions culminated with the 1935 high-latitude *Sadko* expedition led by Zubov (Zubov, 1939) and the 1937 first ice-drift expedition of Ivan Papanin's team of four, 'North Pole-1'. Since the mid-1930s Zubov became a leading expert at the Interdepartmental Bureau on Ice Prognosis under the leadership of Otto Schmidt, the Soviet polar hero known as 'Commissar of the Ice'. In 1938 he summarized the numerous Soviet observations on sea ice made by that time in the book *Marine Waters and Ice* (*Morskie vody i l'dy*), part of which later was developed into the seminal *Arctic Ice* (*L'dy Arktiki*) in 1945.

The Russian sea-ice research was, although intense and in most respects leading, also tightly connected in networks with roots in the key geophysical research centers abroad. One of those was Bergen, which in the interwar period was already a well-established center for oceanography and geophysical research, perhaps most well-known for the Bergen school of meteorology, founded by Vilhelm Bjerknes (Friedman, 1989). Bergen was the point of departure not only of Zubov's methodological development, but also of Hans Ahlmann's long glaciological career in the 1910s and 1920s. In the 1930s, Ahlmann, the leading Swedish glaciologist, established long-term connections with the Soviet scientists, not only glaciologists but first and foremost with meteorologists and oceanographers who studied the ice regime of the Arctic Ocean, like Zubov and Wiese.[13] Ahlmann was to work hand in hand with Harald Ulrik Sverdrup and, as we shall see, Ahlmann's work was also important to US sea-ice research during the Cold War.

Western responses between the wars

The Soviet knowledge of sea ice was not confined to theoretical understanding; new sea-ice technologies were also developed, such as icebreakers, leading the world in size and technological prowess. One example was the *Krassin*, which assisted in the search-and-rescue operations after Umberto Nobile's fatal Zeppelin expedition of 1928 and brought goodwill to Soviet science in a period when the artistic and literary avant-garde all seemed to contribute towards giving a favorable image of communism in Europe. Sea-ice research elsewhere was smaller in scale, built more on individual initiative and lacking in infrastructure.

However, Norway was an exception with her distinct polar ambitions dating back to the work by Nansen on Greenland, the first *Fram* expedition 1895 to 1897 – a veritable passage in drifting sea ice across the Arctic basin – and Amundsen's nationalist claim of the South Pole for Norway. After full independence in 1905 from the union with Sweden, Arctic and Antarctic issues were at the core of foreign affairs for Norway, a whaling, sealing, fishing nation with strong traditions in minerals and energy (Berg, 1995). Norway's Arctic efforts were strongly underpinned by the preparations for the Svalbard Treaty, negotiated in 1920, in which Norway was given a primary role as the governing state with de facto sovereignty, although all signatory states had the right of access to resources and economic use. Building a Norwegian North Atlantic presence was an impetus to gain natural knowledge in the north, including about sea ice. Thus sea ice was a major feature of the research program of the *Maud* expedition, 1918 to 1925, that was supposed to pursue the transect of the Arctic Ocean attempted by Nansen's *Fram* expedition but never realized. The expedition was conceived of by Roald Amundsen in 1908 but was postponed, first by Peary's discovery of the North Pole in 1909, then by Amundsen's expedition to the South Pole, and finally by World War I. Most of the time the *Maud* was frozen into sea ice that did not drift much and the last three winters of 1923, 1924, and 1925 she was stuck near the coast in Eastern Siberia.

During this stretch of the expedition the young Swedish meteorologist Finn Malmgren was put in charge of a special investigation of sea ice, along with his collection of meteorological data and studies of frost and cloud formation. Malmgren's research on sea ice was probably the most ambitious work on the element as yet undertaken. He could build on three years of field data collection followed by two years of further preparations until his 67-page paper was ultimately published, also as his dissertation (Malmgren, 1927).[14] Malmgren's interest was almost exclusively on the formation of sea ice: how it froze, its salinity, density, heat-expansion, conductivity, temperature, and so on. There is little evidence that he paid any attention at all to climate changes and their possible effects on the prevalence and long-term fluctuations of sea ice. Nor did he observe any such interest in the work of his many predecessors, for example Nansen, the Russian commander Stepan Makarov, who published a study from his investigations on the *Yermak* icebreaker (Makarov, 1901), or the Swede Axel Hamberg (1895), later a renowned glacial researcher.

However, at the end of his study Malmgren devoted a number of pages to 'Conclusions regarding heat transport to the air from below.' In this section he presented, based on his own data and measurements, the quite sensational idea that considerable heat is transported from the sea through the ice to the air mass above. Not only did this heat conduction from below contribute to the melting of the ice, it also warmed the thin layer of cold air, some 150 meters above sea level, that covered the polar ocean. This last phenomenon had been observed by the expedition leader Harald Ulrik Sverdrup during the same expedition (Sverdrup, 1926). Malmgren himself seemed quite overwhelmed by his finding: 'The effect of the heat conducted from below "– – –" on the temperature is very great – – – This quantity [the average of heat transported from below every day through the winter, September to April] would be sufficient to raise the temperature of the atmosphere in the lowest 150 meters by 6.9° in one day' (Malmgren, 1927, p. 66). This was the explanation, he argued, of the fact that the winter air above the polar sea was so comparatively mild in relation to the temperatures on Arctic continents further south.

Cold War environmental science

Thanks to Soviet and Norwegian work, the level of knowledge in 1939 was far higher than in, say, 1918, although British, French, Danes, Americans and Canadians had also provided significant contributions.[15] After 1945 the political situation in the Arctic changed dramatically and a totally new phase in the formation of sea-ice knowledge started, however with a remarkable persistence of the old trope: the scope of drastic change in ice cover and the possibility of a past, and future, ice-free Arctic sea. For strategic reasons sea ice became important and a grave concern for NATO states during the Cold War, and efforts were made to translate seminal Soviet works into English. Thus Zubov's books were translated, at first in Canada,[16] later in the United States (Zubov, 1963). Earlier a summary of a review paper on Soviet sea-ice research was published in *Polar Record* (Laktionov, 1945). British geographer Terence Armstrong, who was based at the Scott Polar Research Institute in Cambridge, played an important role as mediator in the transfer of knowledge about Soviet Arctic activities. He was fluent in Russian and followed all the publications he could find while doing research for his book, *The Northern Sea Route* (Armstrong, 1952). An overview of sea-ice studies (Armstrong, 1954), and another study specifically on Soviet sea-ice forecasting (Armstrong, 1955) followed. He stressed that the Soviets were leading and that 'North American progress in this field appeared in 1954 to have reached the level achieved in the USSR in 1939' (Armstrong, 1954, pp. 203–4).

The overarching motivation was military and had to do with the then seemingly likely prospect of the Arctic as a theatre of war if armed conflict occurred between the United States/NATO and the Soviet Union. A war in the far north would require that sea ice was used for transportation, storage, and possibly as landing space for special aircraft or helicopters. The thickness of the ice would determine the operation of submarines. The annual and seasonal behavior of sea ice would set limits to naval operations and give hints to the use and deployment of icebreakers. Connections between sea ice and land would determine the mobility of troops. The properties of sea ice, its texture and solidity, would determine how it could be used. Evidently, if there was climate change – already an established fact, in particular with the distinct warming trend in the Arctic between 1918 and 1946, and the cooling trend from then on to the mid-1960s – this would affect sea ice in all possible ways, but how and to what extent?

This was the strategic situation when the Americans started to use Greenland as a platform, first for its World War II operations and on a much increased scale during the Cold War. The United States knew all too well that the Soviet Union had performed advanced sea-ice research in the Arctic for decades. As was pointed out in a secret CIA memo in April 1960, 'environmental research', in particular in the Arctic, had long been undertaken by Soviet scientists 'as an element in the power struggle as compared with that of the Free World', and to give the USSR 'a time advantage in the development of a capacity to forecast or prognosticate the occurrence of physical environmental phenomena, whether for economic competition or for military operations.' The document details the timeline of this research: in 1921 Lenin issued a decree to found the 'Floating Marine Scientific Institute'; in 1932 the new plan to open up the Northern Sea Route and develop the Arctic region became a major objective; in 1937 the NP–1 (acronym for Papanin's ice-floe expedition 'North Pole – 1', or SP-1 for *Severnyi Polyus* in Russian) demonstrated the value of drift stations. Furthermore, the Soviets used aircraft as flying laboratories from which weather and ice observations were made from the coast northward into the interior of the Arctic basin. Instruments had been developed for weather and ice observations installed on icebreakers, 'some being detailed explicitly for studies of the pack ice' (CIA, 1960, pp. 12–13). By 1954 the Soviets already had in place 'a comprehensive range of programs' at a routine level, encompassing polar stations, drift stations, aerial ice-reconnaissance surveys that operated year round, and additional hydrographic, oceanographic, geologic, and glaciological expeditions that operated on a seasonal basis (pp. 13–14). Evidently, the Soviet Union knew more than anyone else about sea ice, a strategic supremacy that the Americans needed to balance.[17]

Thus when in the fall of 1956 the US National Academy of Sciences decided to organize an international conference on sea-ice forecasting (which later, when the theme had been broadened, was renamed Arctic Sea Ice Conference), funded by the Office of Naval Research, the organizers were confident that they should try to persuade Soviet scientists to attend. The possibilities for that had been explored since the summer of 1957, while the conference itself was scheduled for February 1958.[18] Among six chairs of the sessions there was one Soviet scientist, Grigory Avsyuk, head of Soviet glaciological research during the International Geophysical Year (IGY), who chaired the session on sea-ice formation, growth and disintegration.[19] Sverdrup, since the 1920s one of the leading authorities on sea-ice, had agreed to chair a session on forecasting, but died unexpectedly in August 1957. The session 'Distribution and character of sea ice' was led by Terence Armstrong. Among Soviet participants in addition to Avsyuk were leading Arctic scientists such as Pavel Gordienko, who participated in many High Latitude Air Expeditions conducted in the Soviet Union almost every year from 1948 to 1993 (Gordienko, 1958). Due to the work of Gordienko and many of his colleagues the Arctic and Antarctic Research Institute became one of the World Data Centers on sea ice, keeping data since 1930s, including 10-day maps of sea-ice distribution for the period 1933–1992 (Frolov et al., 2012).

In the mounting Cold War the United States sought collaboration with Canadian authorities, which was natural since their northern ally possessed much sea ice and also had more concrete experience of this strange material. The powerful Joint Research and Development Board (JRDB), directly advising the Pentagon and the Joint Chiefs of Staff, put immediate attention to questions of sea and ice after the war. One important source of inspiration was new knowledge of 'Polar Warming' brought forth by Swedish glaciologist Hans Ahlmann (Sörlin, 2011). On 16 June 1947 Ahlmann appeared at a meeting of the JRDB's Committee on Geophysical Sciences, an event preceded by an initiative in the committee to make use of this new knowledge 'in view of the present emphasis on Arctic warfare' and the fact that 'No research of comparable scientific importance in this field is being done in the United States at the present time.'[20] The response in the committee, and more precisely its Panel on Meteorology where Ahlmann's fellow Swede, the meteorologist Carl-Gustaf Rossby was a member, was to alert the 'Secretaries of War and Navy of the existence of a serious gap in the present research programs of the services' and to recommend that

> the Army and Navy consider assigning qualified personnel to study under Professor Ahlmann. The Panel is of the opinion that such action would assist in filling a definite gap in the research program for national defense.[21]

Clearly Ahlmann's work was chiefly on terrestrial glaciers, although he did also take an interest in sea ice and, perhaps more importantly, he had for a long time argued that there was a possible connection between the behavior of these two basic ice forms. In a lecture to the Annual Meeting of the Royal Academy of Sciences in Stockholm in March 1943, Ahlmann presented a general overview of 'Ice and Sea in the Arctic', and he proposed that it was the same 'influence of the general circulation' that explained first of all the general Arctic warming trend and then both the reduction of glaciers and the extent of sea ice. Thus, the study of glaciers and of general atmospheric circulation in the polar region would generate results of direct relevance for the understanding of sea ice and for predictions of its further behavior. 'Land ice and sea ice...speak the same language.'[22] As interesting as this link to Scandinavian research is the fact that a copy of the translation to English was sent from Rossby's office in February 1947 to the Office of Naval Research, explaining the chain of events that followed in the JRDB during the spring and summer. Clearly this was not the only source of inspiration for the US military to develop work on sea ice but it was an early and very decisive link to the issue of climate change. The chief advantage to be gained was to understand the full extent of the connections between atmospheric factors and sea ice, and thus be able to predict sea-ice conditions necessary for future warfare.[23]

In August 1948 the discussion in the JRDB's Committee on the Geophysical Sciences centered around the assigning of primary responsibility for 'a program of basic and applied research on snow, land-ice, sea-ice, and permafrost.'[24] By January 1949 a special Panel on Arctic Environments had been formed and one of the items of its first meeting was a 'Report of Working Group on Sea Ice', under the Panel on Oceanography.[25] A few months later results seemed so promising that it was suggested that some single agency should be charged with overall responsibility of the ice phenomenon, in order to be able to forecast ice behavior on the basis of meteorological conditions, plus observe ice conditions and develop accurate meteorological forecasting techniques.[26]

In the following years the buildup of ice knowledge was rapid. By 1948 the US Navy began charting sea-ice conditions from reconnaissance flights and in 1951 the Navy began a sustained ice-observing and -forecasting program (*Report*, 1958, p. 2). Assistance from the Canadians was organized chiefly through the head of the Canadian Defense Research Board, Graham Rowley, who oversaw the multiple Canadian facilities for sea-ice monitoring and measurement. He was also among those who predicted an open polar sea and from this he drew the conclusion that such a development would in turn cause more climate change. By 1952 and during the following years new knowledge had started to accumulate and the Hydrographic Office of the United States was issuing reports on the ice conditions in different parts of the North American Arctic.[27] One of the reports was *A Functional Glossary of Ice Terminology*, a compilation based on the scattered but quite comprehensive literature on terminology and classification of ice. It had its roots in a hastily assembled sea-ice manual presented for the operational season of the summer of 1948, but was now enlarged with land-ice terms to cover all polar operational needs, including in Antarctica.

Conclusion: media and the power of narrative

The twentieth-century history of the science politics of sea ice brings a familiar pattern: interest drives science. Although ambitions to build knowledge existed in several countries, it was economic, strategic, military and geopolitical factors that determined who conducted the most systematic efforts. As the geopolitical situation changed, the range of actors did as well. Norway and Russia/Soviet Union provided early efforts; during the Cold War, the United States and Canada enhanced their presence. As Cold War tension in the Arctic eased from the late 1960s and certainly after 1989, research priorities to some extent reverted to the older pattern of the pre–Cold War era. What has come up in more recent years is a growing interest in the relations between sea ice and economic development based on fossil fuels, shipping and strategic minerals, still with the same states involved. Yet research programs today tend to be more multinational and continuous rather than national and expedition-based.

In another sense, the story of sea ice has repeatedly hinted at a pattern that is less familiar. It is research that is conducted in close relation to narratives and tropes about climate change. What is often repeated in these narratives is that the sea-ice cover is highly variable and fairly often it is pointed out that there is a possibility of an ice-free Arctic sea, both at times in the past and in the not-too-distant future. Arctic sea ice is a substance which is like no other – it comes and goes. In contrast to glaciers that act decisively but slowly and predictably on climate change, sea ice is a fugitive element. This is because of its multiple drivers – ocean currents, heat conductivity from below, air pressures, even weather phenomena such as storms and seasonally prevailing winds – that move and change by the week, day and hour, although there are also more stable long-term trends, for example, it seems, those connected to anthropogenic climate change which is bound to affect Arctic sea-ice cover on some timescale.

The media interest in the ever more frequent reports from the National Snow and Ice Data Center, NSIDC, in Boulder, Colorado, on the changes of Arctic sea ice indicates that there is something more than just cryoscience that drives our curiosity. Sea ice, being this highly fluid material, rich in amplitude, seems to capture a contemporary preoccupation with global concerns with the fate of our planet. It is increasingly interpreted as a global indicator of how 'we' – humanity at large – impact on the world, at the same time as the waning of the sea ice is visible and tangible proof of the prospect of a new economic order in the far north, one that both lures and scares.

Central to this stands the trope of an ice-free Arctic sea. Although some now cherish this idea for economic reasons, a common trait in almost all comments about it is that it is an anomaly, perhaps even a monstrous phenomenon. The idea that there might at some point have existed an ice-free Arctic in the Holocene past was proposed during the nineteenth and twentieth centuries based on then-current science.[28] Albeit mostly regarded as phantasmal, almost mythical, and thus incredible, it is important because it tends to sustain a narrative of a certain cyclical pattern of the Earth, of a flux between states, thus perhaps providing some confidence in the otherwise scary notion that we might be on our way towards that open polar sea again. The symmetrical but much later trope of a *future* open polar sea – dating in its current form from the immediate postwar years in the late 1940s – is equally science-based. But there is a fundamental difference. When the older trope suggested a return to a future ice-free state it did so based on major, although as yet unrevealed, properties of the Earth, the atmosphere, and possibly extra-atmospheric forces. 'It is tempting', Hans Ahlmann finally concluded in his seminal 1943 paper on 'Sea and Ice in the Arctic', 'to look for the cause in the change of the amount of heat which is supplied to the earth by the sun, but there are no proofs of this as yet.' No, there surely was no proof of that. It was very big, still unknown but in principle intelligible, forces that drove the Arctic and its ice back and forth between extremes.

The new trope for a long time had no major audience. It was a passing thing. True, major media reported on it in the first half of the 1950s. The *New York Times* wrote as the Soviet Union was joining the International Geophysical Year 1957/58 in December 1954 that there was likely a global warming that might significantly affect the Arctic: 'How rapidly is the earth warming up? There are indications that the rate of warming has accelerated since 1900 and that in twenty-five to fifty years the ice may melt out of the Arctic Ocean in the summer.'[29] Scientist Paul A. Siple, a Cold War glory boy who hit the cover of *Time* magazine, made precisely this prediction in 1953.[30] But all in all the trope of a future open Arctic sea by 2000 fell into oblivion. When now resurrected, although placed sometime in the 2030s or so, it is not necessarily because we are more, or less, concerned. The difference is that there has been a major shift in both the narrative and the character of the science that is behind it. Still, for Canadian sea-ice specialist Graham Rowley, American Paul Siple, and others who talked of a soon-to-be-open Arctic sea in the early 1950s, the forces at play were natural, not anthropogenic. When this changed, and humans were implied, the entire ethics and politics of the trope changed too.

One could see this as a change that occurred in two steps. The immediate postwar years were a breakthrough for what has been called 'the environmental' geophysical sciences, a qualitatively new discourse of global change and of man–earth interactions that put more and more responsibility on the human role in changing the Earth (Robin, Sörlin and Warde, 2013). The environment stood at the core of the US military's research on geophysical factors, including sea ice.[31]

Only gradually, however, did the idea of human-induced climate change enter into the environmental discourse, and when it did, aided by the emergence of computer-based atmospheric circulation models, sea-ice was for a long time not included. When it was, starting in the 1970s and based on new insights of the critical importance of sea-ice albedo feedbacks, it turned out to become very important both as a trigger of large-scale climate change and as a potential controller of ocean circulation.[32] It was not until around 2000 that this new situation was canonized under the captive concept of the Anthropocene, coined by atmospheric scientist Paul Crutzen (Crutzen and Stoermer, 2000). Under this new narrative, the trope of the open Arctic sea seems to have flipped from a mythical feature of the distant past to an alarming signal from the future: this is what we humans will achieve, an ice-free Arctic Sea with oil rigs on the shores, a telling metaphor of human destructive foolhardiness and hypocrisy.

This, in essence, is what is behind the preoccupation of the media around the world when they tell us, year after year in the late summer and early fall, of a dotted downward line in the graphics on the NSIDC website: sea-ice levels in the Arctic approaching yet another minimum? Whether there will ever be a seasonally open Arctic Ocean we don't know. Whether there has ever been one, we, in earnest, don't know either. That we will continue to pay attention, we do know.

Acknowledgments

We are deeply indebted during many years of collaboration under the European Science Foundation's BOREAS program to Ron Doel, especially for allowing us to use previously unpublished material from American archives. Our thanks also extend to Peder Roberts who has agreed to share archival materials from the National Academies Archives in Washington, DC. We are also grateful to the editors, our co-contributors to this volume for rewarding discussions and comments, and to several reviewers. Lajus gratefully acknowledges support from the Basic Research Program of the National Research University Higher School of Economics, St Petersburg in 2012 for the project 'Circulation of Knowledge in a Divided World: Attraction, Confrontation, Cooperation among Communities of Experts in the Cold War Period', and support from the Swedish Mistra Foundation's project, 'Assessing Arctic Futures'. Sörlin acknowledges support from the Swedish Mistra Foundation's project, 'Assessing Arctic Futures', and to the Swedish Research Council for Environment, Agricultural Sciences and Spatial Planning (Formas) for the project 'Models, Media, and Arctic Climate Change'.

Notes

1. http://www.nsidc.gov, accessed 7 November 2011.
2. An early usage of the word 'environmental' in its modern sense is by William Vogt in *Road to Survival* (1948) (London: Victor Gollancz, 1949), pp. 14–15. S. Sörlin and P. Warde (2009), 'Making the environment historical – an introduction' in *Nature's End: History and the Environment*, ed. P. Warde and S. Sörlin (London: Palgrave Macmillan), pp. 1–19. R. Grove and V. Damodaran, 'Imperialism, Intellectual Networks, and Environmental Change: Unearthing the Origins and Evolution of Global Environmental History' in Warde and Sörlin, *Nature's End*, pp. 23–49.
3. The word itself derives from the Ancient Greek word 'κρύος' (*cryos* meaning 'cold', 'frost' or 'ice'). It is the term which collectively describes the portions of the Earth's surface where water is in solid form, including sea ice, lake ice, river ice, snow cover, glaciers, ice caps and ice sheets, and frozen ground (which includes permafrost). http://en.wikipedia.org/wiki/Cryosphere, accessed 18 February 2013.
4. There was between 1.5 and 3.0 degrees' warming from 1920 to 1940 based on observations from several Arctic and North Atlantic stations; the warming was especially pronounced in Spetsbergen where winter temperatures in the 1930s increased up to 10 degrees Celsius. See further: http://www.arctic-warming.com.
5. The entire volume, Krupnik et al. (2010), *SIKU: Knowing Our Ice*, bears evidence to this; see especially Chapter 1, I. Krupnik, C. Aporta and G. J. Laidler, 'SIKU: International Polar Year Project #166 (An Overview)', pp. 1–30.
6. M. T. Bravo, 'The Humanism of Sea Ice', in Krupnik et al. (2010), pp. 447–8.
7. L. Koch (1945), *The East Greenland Ice*. Meddelelser om Grönland, 130:3 (Copenhagen: Reitzel). A. Kolchak (1928) 'The Arctic Pack and the Polynya', *Problems of Polar Research* (New York: American Geographical Society), pp. 125–41. (This was a posthumous publication of a chapter from his seminal book *"L'dy Karskogo i Sibirskogo morei" [Ices of Kara and Siberian Seas*, St. Petersburg, 1909] which was under full prohibition in the Soviet time due to obvious political reasons – Kolchak was a Supreme Ruler of Russia during the Russian Civil War in 1918–1920 and was killed by Bolsheviks). A. Maurstad (1935), *Atlas of Sea Ice*, Geofysiske publikasjoner. 10:11 (Oslo: Cammermeyers Boghandel distr.). E. H. Smith (1932) 'Ice in the Sea', Chapter 10 of *Physics of the Earth V: Oceanography* (Washington: The National Research Council), pp. 384–408, a very rich and comprehensive review without any mention at

all of changes in climate other than the 'several glacial periods' (p. 384). N. A. Transehe (1928), 'The Ice Cover of the Arctic Sea, with a Genetic Classification of Sea Ice', *Problems of Polar Research* (New York: American Geographical Society), pp. 91–123.

8. For the general history of *Glavsevmorput*, see J. McCannon (1998) *Red Arctic: Polar Exploration and the Myth of the North in the Soviet Union, 1932–1939* (New York: Oxford University Press), especially pp. 33–58.
9. This paper was also published in German in a Swedish geography journal: V. Wiese (1924), 'Polareis und atmosphärische Schwankungen', *Geografiska Annaler* VI, 3:4, 273–99.
10. This work is still ongoing and recent publications confirm the downward trend since the late nineteenth century. K. Lassen (1997) Twentieth century retreat of sea-ice in the Greenland sea, DMI Sci. Rep. pp. 95–7.
11. See Sverdrup's archives in The National Library, Oslo (Harald Ulrik Sverdrup, Vitenskapelig arkiv, Brevsamling N 634). The earliest letter from Wiese in this collection is from 1927, however, it does not look like a first letter but as a piece from continuous correspondence. Several books by Sverdrup including materials from the "Maud" expedition (1918–1925) with Roald Amundsen and description of the "Nautilus" submarine expedition with George Hubert Wilkins (1931) were translated into Russian in 1930 and 1932.
12. Zubov (1933) p. 400, reference to Ahlmann (1932): 'L'expédition arctique suédo-norvégienne (Terre du Nord-Est et mers voisines)', *Annales de Géographie*, 41:230, 177–87. Ahlmann was arguably the leading glaciologist at the time with a considerable reputation both in Europe, North America and increasingly among Soviet colleagues. See J. Lajus and S. Sörlin, 'Melting the Glacial Curtain: Scandinavian–Soviet networks in the geophysical field sciences between two Polar Years, 1932/33–1957/58', *Journal of Historical Geography* (in press).
13. Lajus and Sörlin, 'Melting the Glacial Curtain'.
14. This work by Malmgren was translated into Russian and published in 1930 with Malmgren's obituary written by H. U. Sverdrup; see F. Malmgren (1930), *O svoistvakh morskogo l'da* (Leningrad: Gosudarstvennyi gidrodraficheskii institut).
15. A notable American example is E. H. Smith (1931), *The Marion Expedition to Davis Street and Baffin Bay under the Direction of the U.S. Coast Guard 1928*, Scientific Results Bulletin, 19, Part 3 (Washington, DC). See also the country by country bibliography in Transehe (1928), 'The Ice Cover'.
16. See The Papers of N. N. Zubov in the Dartmouth College Library. Location: Baker Spec. Coll., Manuscript Stef.Mss/106 http://ead.dartmouth.edu/html/stem106.html. There are translations of some of his works done in 1947, among them the translation of the book *In the Center of the Arctic* published as *V tsentr Arktiki: ocherki poistorii issledovaniia I fizicheskoi geografii tsentral'noi Arktiki* (Moskva: Izdatel'stvo Glavsevmorputi, 1948) which contains a chapter 'Warming of the Arctic'. The translation was edited by Vilhjamur Stefansson. It was published in 1950 in Ottawa by the Defense Research Board. The same is true for the following: N. N. Zubov's (1948), *Arctic ice and the warming of the Arctic, being chapters VI and VII of the center of the Arctic; an outline of the history of Arctic exploration and of the physical geography of the central Arctic* (Northern Sea Route Directorate Press, Moscow-Leningrad), trans. E. Hope (1950) (Ottawa: Defense Research Board).

17. 'Annex B: Environmental Research and Its Implications to Long-Range Power Positions', 26 April 1960 (subsequent draft of initial mid-1950s analysis), CIA-RDP63–0031R0002000170005–9, CIA CREST, NARA II, Washington, DC (CIA 1960). General on the Soviet drifting stations, SP-1 through SP-7, in T. Armstrong (1958), *The Russians in the Arctic: Aspects of Soviet Exploration and Exploitation of the Far North, 1937–57* (London: Methuen), pp. 67–79.
18. See documents from the National Academies Archives, Central Policy Files, 1957–1961. International Relations: International Congresses: Sea Ice Conference, 1957. The authors are extremely grateful to Peder Roberts who shared these documents with them.
19. *Arctic Sea Ice. Proceedings of the Conference conducted by the Division of Earth Sciences and supported by the Office of Naval Research/conference held in Easton, Maryland, February 24–27, 1958*. Publication 598 (Washington, DC: National Academy of Sciences – National Research Council, December 1958), Preface.
20. Remarks by Professor Hans Ahlmann of the University of Stockholm, Sweden, Appendix A to minutes, JDRB's Committee on Geographical Exploration, 16 June, Box 227, folder18, Entry 341, RG 330 (Research and Development Board), National Archives (II), College Park, MD. For the background to Ahlmann's appearance in the JRDB, see Sverker Sörlin (2009), 'Narratives and Counter Narratives of Climate Change: North Atlantic Glaciology and Meteorology, ca 1930–1955', *Journal of Historical Geography*, 35:2, 237–55.
21. Memo by Rossby 21 April 1947, and by C. S. Piggot, Executive Director of the Committee on Geophysical Sciences 12 June 1947, both to the JRDB Executive Secretary, including draft letter by Vannevar Bush, in Box 162, folder 9, Entry 341, RG 330 (Research and Development Board), National Archives (II), College Park, MD.
22. Hans Wilhelmsson Ahlmann, 'Ice and Sea in the Arctic' (orig. in Swedish), Lecture at the Annual Meeting of the Royal Academy of Sciences in Stockholm, 31 March 1943. Copy of English translation in the Vilhjalmur Stefansson collection, Dartmouth College, Stef. MSS-242 (1). Ahlmann (1943), 'Is och hav i Arktis', *K. Svenska Vetenskapsakademiens Årsbok* (Stockholm: Almqvist & Wiksell), 327–336.
23. See cover letter to Capt. Howard Hutchinson in the Upper Atmosphere Panel, Office of Naval Research, by office manager Ethel Pearson in Rossby's Department of Meteorology at the University of Chicago, 19 February 1947: 'Dr. Rossby asked me to send [Ahlmann's paper] to you.' Vilhjalmur Stefansson collection, Dartmouth College, Stef. MSS-242 (1).
24. Meeting 1948–08–30: RG 330, Entry 342, B. 20, x,RDB_Transcripts_file_2_#60.
25. Meeting 1949–01–04: RG 330, Entry 341,169,2,x,RDB_file_3_#59.
26. 19490404 RG 330, Entry 341, 169,18,x,RDB_file_3_#59.
27. For example, reports HO 552 *Ice conditions on Baffin Bay and the Canadian Arctic, Summer 1951: Operations Blue Jay and Nanook* (1952); and HO 553 *Oceanographic data for Davis Straight and the Labrador Sea* (1952).
28. The trope was in fact much older, traceable in Renaissance sources and continued in navigators' and geographical lore, but gained a much more scientific foundation in nineteenth- and early twentieth-century sources. A splendid overview of these truly fabulous discussions can be found in John K. Wright (1953), 'The Open Polar Sea', *Geographical Review* 43, pp. 338–65. It is

clear that Wright was writing from a position in time and scientific knowledge where the trope of an ice-free Arctic could be judged as 'crackpot' (p. 364); little would he know that the idea has now gained new currency based on recent researches; see Funder and Kjaer (2007) and Polyak et al. (2010).
29. 'Soviet Joins 1957–58 World Research', *New York Times*, 11 December 1954.
30. Paul Allman Siple's iconic portrait, with fur-brimmed hat, and suitably staged with iceberg, helicopter, surveillance aircraft, and trodding penguins, was on the *Time* cover 31 December 1956. Siple was the epic man of cold, he coined the concept 'wind-chill factor', went six times to Antarctica, and worked for most of his career as a military scientist, embodying the relationship of environmental science, big-scale prediction and national security.
31. This point has been most forcefully made by Ron Doel in a string of works over the last decade; see for example R. E. Doel (2003), 'Constituting the Postwar Earth Sciences: The Military's Influence on the Environmental Sciences in the USA after 1945', *Social Studies of Science* 33:5, pp. 635–66.
32. This change in climate modeling was originally spurred by articles modeling the behavior and wider role of sea-ice by Mikhail Budyko and William Sellers in the late 1960s. See M. T. Greene (2007), 'Arctic Sea Ice, Oceanography, and Climate Models,' in *Extremes: Oceanography's Adventures at the Poles* (eds), K. R. Benson and H. M. Rozwadowski (Sagamore Beach, MA: Watson International Publishing, 303–329).

References

Ahlmann, H. (1932) 'L'expédition arctique suédo-norvégienne (Terre du Nord-Est et mers voisines)', *Annales de Géographie*, 41:230, 177–87.
—— (1943), 'Is och hav i Arktis', *Kungl. Svenska Vetenskapsakademiens Årsbok* (Stockholm: Almqvist & Wiksell), 327–36.
Arctic Sea Ice. Proceedings of the Conference conducted by the Division of Earth Sciences and supported by the Office of Naval Research/conference held in Easton, Maryland, February 24–27, 1958. Publication 598 (Washington, DC: National Academy of Sciences – National Research Council, December 1958, Preface).
Armstrong, T. (1952) *The Northern Sea Route*. Scott Polar Research Institute Special Publication N 1 (Cambridge: Cambridge University Press).
—— (1954) 'Sea Ice Studies', *Arctic*, 7:3–4, 201–5.
—— (1955) 'Soviet Work on Sea Ice Forecasting', *Polar Record*, 7:49, 302–11.
—— (1958) *The Russians in the Arctic: Aspects of Soviet Exploration and Exploitation of the Far North, 1937–57* (London: Methuen).
Bader, H. (1949) 'Trends in Glaciology in Europe', *Geological Society of America Bulletin*, 60:9, 1309–14.
Bent, S. (1872) *Thermal Pathways to the Pole* (St. Louis: R.P. Studley Co).
Berg, R. (1995) *Norsk utenrikspolitiks historie* (Oslo: Universitetsforlaget).
Brooks, C. E. P. (1925) 'The Problem of Mild Polar Climates', *Quarterly Journal of the Royal Meteorological Society*, 51, 83–94.
—— (1949 [1926]) *Climate Through the Ages*, 2nd rev. ed. (New York: Dover).
CIA (1960) 'Annex B: Environmental Research and its Implications to Long-Range Power Positions', 26 April 1960 [subsequent draft of initial mid-1950s analysis], CIA-RDP63–0031R0002000170005–9, CIA CREST, NARA II, Washington, DC.

Clague, J., and O. Slaymaker (2000) 'Canadian Geomorphology 2000: Introduction', *Geomorphology*, 32:3–4, 203–11.
Clarke, G. K. C. (1987) 'A Short History of Scientific Investigations on Glaciers', *Journal of Glaciology: Special Issue*, 4–24.
Crutzen, P. J., and E. F. Stoermer (2000) 'The "Anthropocene"', *Global Change Newsletter*, 41, 17–18.
Doel, R. E. (2003) 'Constituting the Postwar Earth Sciences: The Military's Influence on the Environmental Sciences in the USA after 1945', *Social Studies of Science*, 33:5, 635–66.
Fagan, B. (2000) *The Little Ice Age: How Climate Made History, 1300–1850* (New York: Basic Books).
Friedman, R. M. (1989) *Appropriating the Weather: Vilhelm Bjerknes and Construction of a Modern Meteorology* (Ithaca, NY: Cornell University Press).
Frolov, I. E., et al. (2012) 'Morskoi Lyod', in *Metody otsenki posledstsvii inzemeniia klimata dlia fizicheskikh i biologicheskikh sistem* (Moscow: Rosgidromet, pp. 400–29).
Funder, S., and K. Kjær (2007) 'Ice Free Arctic Ocean: An Early Holocene Analogue', *Eos, Transactions of the American Geophysical Union*, 88:52, Fall Meeting Supplement, Abstract PP11A-0203.
Gordienko, P. (1958) 'Arctic Ice Drift', in *Arctic Sea Ice. Proceedings of the Conference conducted by the Division of Earth Sciences and supported by the Office of Naval Research*, conference held in Easton, Maryland, February 24–27, 1958. Publication 598 (Washington, DC: National Academy of Sciences – National Research Council, December 1958, pp. 210–22).
Greene, M. T. (2007) 'Arctic Sea Ice, Oceanography, and Climate Models,' in *Extremes: Oceanography's Adventures at the Poles*, eds. K. R. Benson and H. M. Rozwadowski (Sagamore Beach, MA: Watson International Publishing), pp. 303–29.
Grove, J. (1988) *The Little Ice Age* (London: Routledge).
Grove, R., and V. Damodaran (2009) 'Imperialism, Intellectual Networks, and Environmental Change: Unearthing the Origins and Evolution of Global Environmental History', in S. Sörlin and P. Warde (eds), *Nature's End: History and the Environment* (Basingstoke, UK: Palgrave Macmillan), pp. 23–49).
Hamberg, A. (1895) *Studien über Meereis und Gletschereis*, Bilaga till Kungl. Vetenskapsakademiens Handlingar, 21:2:2 (Stockholm).
Kerner, F. von Marilaun (1910) 'Klimatogenetische Betrachtungen', in W. D. Matthews, *Hypothetical Outlines of the Continents in Tertiary Times*, Verh. k. k. geol. Reichsanst. Wien, 12, pp. 259–84.
Knipowitsch, N. M. (1921) 'O termicheskikh usloviiakh Barentseva moria v kontse maia 1921 goda', Bulleten' Rossiiskogo gidrologicheskogo instituta 9: 10–12.
Koch, L. (1945) *The East Greenland Ice*. Meddelelser om Grönland, 130:3 (Copenhagen: Reitzel).
Kolchak, A. (1928) 'The Arctic Pack and the Polynya', in *Problems of Polar Research* (New York: American Geographical Society), pp. 125–41.
Krupnik, I., et al. (eds) (2010) *SIKU: Knowing Our Ice* (Heidelberg, London and New York: Springer).
Lajus, J., and S. Sörlin, 'Melting the Glacial Curtain', *Journal of Historical Geography* (in press).

Laktionov, A. F. (1945) 'Itogi issledovaniya ledyanogo pokrova morey sovetskoy Arktiki i ledovyye prognozy' (Results of investigations of the ice cover of the seas of the Soviet Arctic and ice forecasts), *Izvestiya Vsesoyuznogo Geograficheskogo Obschestva*, 77:6, 341–50. (English summary in *Polar Record*, 5:39 (1950), 468–71).

Lassen, K. (1997) *Twentieth Century Retreat of Sea-Ice in the Greenland Sea*, DMI Sci. Rep. pp. 97–5.

Lüdecke, C., and J. Lajus (2010) 'Second International Polar Year (1932–1933)', in S. Barr and C. Lüdecke (eds), *The History of the International Polar Years (IPYs)* (Berlin and Heildelberg: Springer).

McCannon, J. (1998) *Red Arctic: Polar Exploration and the Myth of the North in the Soviet Union, 1932–1939* (New York: Oxford University Press).

Makarov, S., (1901) *'Yermak' vo l'dakh* [The 'Yermak' in the Icefields] (St. Petersburg).

Malmgren, F. (1927) 'On the Properties of Sea-Ice', in H. U. Sverdrup (ed.), *Norwegian North Polar Expedition with the 'Maud' 1918–1925, Scientific Results* 1:5.

—— (1930) *O svoistvakh morskogo l'da* (Leningrad: Gosudarstvennyi gidrodraficheskii institut).

Matthes, F. E. (1939) 'Report of the Committee on Glaciers', *Transactions of the American Geophysical Union* 20, 518–23.

Maurstad, A. (1935) *Atlas of Sea Ice*, Geofysiske publikasjoner. 10:11 (Oslo: Cammermeyers Boghandel distr.).

Nansen, F., and B. Helland-Hansen (1909) *The Norwegian Sea* (Christiania).

Overland, J. E., K. R. Wood, and M. Wang (2011) 'Warm Arctic – Cold Continents: Climate Impacts of the Newly Open Arctic Sea', *Polar Research*, 30, 15787, DOI: 10.3402/polar.v30i0.15787

Pettersson, O. (1905) 'On the Probable Occurrence in the Atlantic Current of Variations Periodical, and Otherwise, and Their Bearing on Meteorological and Biological Phenomena', *Rapports et Procès-Verbaux des Réunions de Conseil Permanent International pour l'Exploration de la Mer*, 42, 221–40.

—— (1914) 'Climatic Variations in Historic and Prehistoric Time', *Svenska hydrogr. biol. ommissionens skrifter*, 5, 26 pp.

—— (1915) 'Long Periodical Variations of the Tide Generating Force', *Publication Circular Conseil Permanent International pour l'Exploration de la Mer*, 65, 2–23.

Polyak, L., et. al. (2010) 'History of Sea Ice in the Arctic', *Quaternary Science Reviews*, 29, 1757–78.

Report of the Ice Observing and Forecasting Program, 1958 (1958) (Washington, DC: U.S. Navy Hydrographic Office).

Robin, L., and W. Steffen (2007) 'History for the Anthropocene', *History Compass*, 5:5, 1694–1719.

Robin, L., S. Sörlin and P. Warde (2013) 'Introduction: Documenting Global Change', in L. Robin, S. Sörlin and P. Warde (eds), *The Future of Nature: Documents of Global Change* (New Haven, CT: Yale University Press, pp. 1–14.

Rockström, J., et al. (2009) 'Planetary Boundaries: Exploring the Safe Operating Space for Humanity, *Nature* 461, 472–5.

Siple, P. (1953) 'Proposal for Consideration by the U.S. National Committee (UGY)', 1 May 1953, C1, USNC-IGY, National Academy of Sciences, Washington, DC.

Smith, E. H. (1931) *The Marion Expedition to Davis Street and Baffin Bay under the Direction of the U.S. Coast Guard 1928*, Scientific Results Bulletin, 19, Part 3 (Washington DC).
—— (1932) 'Ice in the Sea', Chapter 10 of *Physics of the Earth V: Oceanography* (Washington: The National Research Council, pp. 384–408).
Sörlin, S. (2009) 'Narratives and Counter Narratives of Climate Change: North Atlantic Glaciology and Meteorology, ca. 1930–1955', *Journal of Historical Geography*, 35:2, 237–55.
—— (2011) 'The Anxieties of a Science Diplomat: Field Co-production of Climate Knowledge and the Rise and Fall of Hans Ahlmann's "Polar Warming"', *Osiris* 26: *Revisiting Klima*, J. R. Fleming and V. Jankovich (eds) (Chicago: University of Chicago Press, 66–88).
Sörlin, S. and P. Warde (eds) (2009) *Nature's End: History and the Environment* (London: Palgrave Macmillan).
Steffen W., P. J. Crutzen and J. R. McNeill (2007) 'The Anthropocene: Are Humans Now Overwhelming the Great Forces of Nature?' *Ambio*, 36, 614–21.
Sverdrup, H. U. (1926) 'The North Polar Cover of Cold Air', *Monthly Weather Review*, 53, 471–2.
—— (1931) 'The Drifting Ice', in *The Andrée diaries being the diaries and records of S.A.Andrée, Nils Strindberg and Knut Fraenkel written during their balloon expedition to the North Pole in 1897 and discovered on White Island in 1930, together with a complete record of the expedition and discovery*, transl. by Edward Adams-Ray (London: J. Lane, pp. 255–66).
Transehe, N. A. (1928) 'The Ice Cover of the Arctic Sea, with a Genetic Classification of Sea Ice', *Problems of Polar Research* (New York: American Geographical Society, pp. 91–123).
Treshnikov, A. F. (1978) 'Professor Wiese', in *Ikh imenami nazvany korabli nauki* (Leningrad: Gidrometeoizdat, pp. 4–70).
Utterström, G. (1955) 'Climate Fluctuations and Population Problems in Early Modern Europe', *Scandinavian Economic History Review*, 3, 3–47.
Vogt, W. (1949) *The Road to Survival* (1948) (London: Victor Gollancz).
Wiese, V. (1923) 'O vozmozhnosti predskazaniia sostoianiia l'dov v Barentsevom more', *Izvestiia Tsentral'nogo meteorologicheskogo biuro*, 1, 1–41.
—— (1924) 'L'dy v poliarnykh moriakh i obschaia tsirkuliatsiia atmosfery' [Ice in the polar seas and general circulation of atmosphere], *Zhurnal geofiziki i meteorologii*, 1:1, 78–84.
Wright, J. K. (1953) 'The Open Polar Sea', *Geographical Review*, 43, 338–65.
Zubov, N. N. (1932) *20 dnei na ledovom more (Barentsevo more)* (Moscow: Izdanie Gidrometeorologicheskogo komiteta SSSR i RSFSR).
—— (1933) 'The Circumnavigation of Franz Josef Land', *Geographical Review*, 23:3, 394–401, 528.
—— (1939) *Trudy Pervoi vysokoshirotnoi ekspeditsii na 'Sadko' v 1935 godu*, part 1, vol. 1 (Leningrad).
—— (1963) *Arctic Ice* [L'dy Arktiki], translated by the U.S. Navy Oceanographic Office and the American Meteorological Society (San Diego, CA: U.S. Navy Electronics Laboratory).

5
Signals from a Noisy Region
Annika E. Nilsson and Ralf Döscher

Introduction

When the Arctic Ocean ice cover reached a record low in the late summer of 2007, it provided images of the Arctic entering an era where human-induced climate change had started to create a new geography – the ultimate evidence of the 'Anthropocene' where human actions are a major driver of Earth as a system (Crutzen and Stoermer, 2000). Given that emissions of greenhouse gases continue to increase and that the connection between these emissions and increasing global temperatures is firmly established (IPCC, 2007a), a prominent assumption is that the ice will continue to retreat. In public discussions, the question is no longer *whether* the Arctic Ocean will be virtually ice-free in the summer, but *when* this will happen.

The strong messages of overall *change* of the Arctic Ocean sea ice fit well with general scientific consensus (AMAP, 2011a, 2011b). However, looking beyond headlines and summaries of science reveals a parallel story that places the emphasis on the *variability* of climate in the Arctic. The ice may yet present both science and society at large with new surprises. The focus on change or variability can be described as two different ways of framing Arctic climate change.

The translation of scientific insights about the Arctic climate change into messages for the public and policymakers is by necessity a simplification of the knowledge available. It is the craft of identifying the main and socially most relevant signal in a wealth of scientific information and articulating it in a way that will reach the intended audiences. Such translations take place in social settings that make some signals more relevant than others. The translation of scientific knowledge about Arctic climate change also takes place in a highly political setting, colored by negotiations about limiting the emissions of greenhouse gases and controversies surrounding climate science. This chapter takes a step back from the media and policy messages and attempts to capture the dominant scientific framings of Arctic climate change as visible both in the details of science plans and in scientific assessments and in their summaries for policymakers. Framing refers to how we define a problem, its impacts and potential solutions in ways that highlight certain aspects and downplay others (Mitchell et al., 2006). A certain way of framing an issue or phenomenon guides how the world around us becomes visible (König, 2006). A very dominant framing can leave important issues outside our lines of vision and potentially make society blind to important challenges. Frame analyses are increasingly used in media studies of the climate change discourse (for example Olausson, 2009).

The purpose and rationale for the chapter is twofold. The first is to reveal some basic assumptions that are made about what we can expect from the future regarding the impacts of climate change, and thus need to plan for.

The second rationale is to better understand how translations of scientific knowledge to policy messages are influenced by the social context in which these translations take place. The chapter thereby provides background and analysis of the science and science–policy interactions that have played an important role in shaping media images of Arctic climate change. It is inspired by and provides a contribution to a growing literature about the sociology and history of climate change science in science and technology studies (for example Shackley et al., 1999; Miller and Edwards, 2001; Edwards, 2010).

Understanding how we see the world and what we are blind to is not only of academic interest. In fact, it is relevant to a range of decisions regarding adaptation to climate change. Historically, the Arctic climate has been extremely variable. This is evident both from observations of the physical environment (McBean, 2005) and studies of the dynamics of the Arctic climate system using climate models (Kattsov and Källén, 2005). Moreover, anthropologists studying indigenous people in the Arctic stress that the Arctic environment has been fluctuating ever since the region was colonized by people, and that the archaeological evidence indicates that some of those changes have been as rapid as those predicted for current global warming (Csonka and Schweitzer, 2004). Losing sight of this noise may lead to decisions that are less than helpful in trying to adapt to the long-term changes.

Whether we have or have not lost sight of the noise underlying Arctic climate change is an empirical question. This chapter attempts to answer it by analyzing the framing of Arctic climate change in some texts that are situated at the science–policy interface, as described in more detail below. It also discusses some factors in the scientific and political contexts that may have played a role for shifts in emphasis over time.

The chapter is organized as follows: first are brief introductions to the science and politics of detection and attribution and a further description of the methodology. This is followed by the framing analysis of different science–policy processes in a chronological order, divided into two major periods: before and after 2007. The results are summarized in the concluding discussion, which also presents some reflections on the social implications of the findings.

The science of detection and attribution

One of the most central questions at the interface between climate science and climate policy has been whether the global climate has indeed changed and why. To set the stage for an analysis of the framing of Arctic climate, this section provides a short introduction to the science behind these questions, which in scientific terminology is the science of detection and attribution. Detection is about knowing whether the climate has indeed changed and attribution is about why it has changed. The science relies on a combination of observations of the climate and how it has changed over time, and our understanding of how the climate is likely to vary independently of emissions of greenhouse gases or other external factors. More specifically, detection of climate change is about showing that the climate has changed in a statistical sense, that is, establishing that there is a difference from how the climate has behaved in the past, including the normal variation from year to year, decade to decade, century to century and so on. These 'normal' variations in things like temperature, precipitation and wind patterns are referred to as climate variability.

Historical records of the climate are important for understanding climate variability, but actual observations of temperatures, precipitation and wind patterns do not go back far enough in time to capture all relevant phenomena. Instead it becomes necessary to rely on so-called proxy records. They include tree rings, sea sediments, cores from polar ice caps and other indirect signs of past climates. Interpreting these records in terms of surface temperature has been an important part of establishing a global-scale climate history. The methods of interpretation have long been discussed (Jones and Mann, 2004) and the resulting temperature development curves have been at the center of scientific controversies (von Storch et al., 2009).

The more observations that are available, the easier it is to statistically distinguish a signal of change that goes beyond the normal variability. If the normal variability (not influenced by greenhouse gases) is very large, more observational data in time and space would be required. Climate variability is thus a limiting factor in the detection of trends in that the signal of change that needs to be detected is easily hidden in the noise of a highly variable climate. Lack of data about past variability makes it even more difficult to statistically prove that there is a trend that indicates a more fundamental change. These demands create challenges in assessing whether current Arctic climate change and the declining sea-ice cover are indeed part of the global picture of human-induced climate change, simply because the variability in the Arctic is large and the observations sparse.

One way to better understand climate variability is to use climate models. Climate models are essentially the same type of computer programs that are used for making weather prognosis but they are run over longer time periods. They consist of numerical representations of different physical processes that are involved in creating climate and weather. Modern climate models take into account processes relating to the atmosphere, land surface, ocean and sea ice and also include descriptions of interaction between those components.

To better understand the role of greenhouse gases and other possible causes of climate change, climate models include projections of concentrations of greenhouse gases in the atmosphere and aerosols, along with information about energy input from the sun and a range of other factors that are known to be relevant for Earth's energy balance and thus for the global climate. The output from the models gives insights not only about the potential for climate change under different conditions but also the frequency and character of the natural climate variability.

Attribution of climate change is the process of establishing the most likely causes for a detected change with some level of statistical confidence (Stott et al., 2010). For global climate change, attribution has been possible since the early 1990s (for example Hegerl et al., 1996). Statistical methods were developed to find patterns in the observations that could serve as 'fingerprints' of human influence. This makes it possible to compare patterns from models that are forced with concentrations of greenhouse gases with patterns from actual observations. To achieve statistical significance, results need to be averaged over large areas. The quality of the analysis of attribution benefits from longer observations and from better climate models. Both have been improved since the 1990s and in addition to attributing global climate change to human causes, it has become possible to do similar analyses for individual continents.

For the Arctic, data are still sparse in many areas. We also know from both historical records and climate models that the Arctic climate is highly variable independent of human greenhouse gas emissions. One reason is that the Arctic climate is determined by large-scale, periodic shifts in the atmosphere's circulation linked to powerful high-pressure systems and low-pressure systems. Examples include the Arctic Oscillation and the Arctic Dipole anomaly.[1] In addition, there are positive feedback processes that create local temperature changes, which generate a self-supporting atmospheric circulation (for example Bengtsson et al., 2004). One of the most important is the sea-ice–albedo feedback, where melting sea ice exposes dark water surfaces and thus decreases the amount of solar energy that reflects back into space, which leads to further warming and increasing melt.

For regions such as the Arctic, attribution efforts are dependent on the amount and quality of observational data in the specific region and on the regional amplitude of natural climate variability. Moreover, until recently the statistical tools have also been lacking. In fact, if currently used statistical methods (optimal detection analysis) had been available, human influence on sea-ice loss could actually have been detected as early as 1992 (Min et al., 2008).

The ability to attribute changes in various aspects of the Arctic climate increases when focusing on individual seasons. Arctic temperature change is most detectable in late summer when the sea-ice effect is strongest. Recently, anthropogenic signals have become detectable in colder seasons (ibid.). The 'optimal fingerprint' method, based on a comparison of observations with simulations from a multi-model ensemble, was used for attribution.

Circulation changes in the Arctic atmosphere are more difficult to attribute to human influence. Based on analysis of an ensemble of global climate models, Overland and Wang (2005) concluded that decade-to-decade cycles in the natural variability (such as the Arctic Oscillation) do not rule out the steady influence of greenhouse forcing. Forcing refers to the influence that a factor has in altering the balance of incoming and outgoing energy in the Earth-atmosphere system, in this case the impact of greenhouse gases. Attribution to human influence was supported by a temporary persistent positive phase of the Arctic Oscillation, responsible for a local Arctic warming and sea-ice retreat. Such strong observed signals improve statistical possibilities to detect human influence in comparison with model results.

However, it is difficult to clearly attribute Arctic climate change to human influence based solely on observations (Overland and Wang, 2005). The strategy has therefore been to combine observation-based data with simulations from climate models. In a recent study based on an up-to-date gridded data set of land surface temperatures and simulations from four coupled climate models, Gillet et al. (2008) concluded that anthropogenic influence on Arctic temperature is detectable and distinguishable from the influence of natural forcing, that is, it is directly attributable to human greenhouse gas emissions. They based this on a finding that observed changes in Arctic and Antarctic temperatures are not consistent with internal climate variability or natural climate forcing alone. The conclusion and progress from previous studies was possible due to an updated gridded data set of land temperatures, allowing for more regional comparison with a model ensemble.

The politics of detection and attribution

The international politics of attribution is closely linked to the United Nations Framework Convention on Climate Change (UNFCCC) and its ultimate objective defined as 'stabilization of greenhouse gas concentrations in the atmosphere at a level that would prevent dangerous anthropogenic interference with the climate system' (UNFCCC Article 2). This focus in the international climate regime on anthropogenic interference has made it politically relevant to define to what extent the observed changes are caused by emission of greenhouse gases that have an origin in human activities (anthropogenic) as opposed to any other causes, including natural variability of the climate. The focus on anthropogenic climate change in the UNFCCC has therefore played an important role in scientific assessments of global climate change, including the reports from the Intergovernmental Panel on Climate Change (IPCC). This was particularly visible in debates about wording in the summary for policymakers in the IPCC Second Assessment Report in 1995, which concluded that 'The balance of evidence suggests a discernible human influence on global climate' (Edwards and Schneider, 2001). The evidence that global climate change was indeed caused by human activities became stronger over time, but the debate as such continued to play a role in global climate politics at least until IPCC's fourth assessment in 2007, where Working Group I concluded that 'Most of the observed increase in global average temperatures since the mid-20th century is *very likely* due to the observed increase in anthropogenic greenhouse gas concentrations' (Solomon et al., 2007, p. 10). This is the setting in which the science of Arctic climate change is carried out, in particular the syntheses of scientific knowledge in scientific assessment aimed at supporting policy-makers. The rest of this article presents an analysis of how Arctic climate change is framed in texts aimed at providing a consensus view from science. The analysis forms the basis for a concluding discussion about the possible influence of political context on the scientific discourse and vice versa.

Methodology

In media studies, frame analysis is used as a tool to unmask possibly unintended but nevertheless culturally or politically dependent aspects of news texts (Olausson, 2009). In order to go beyond quantitative word counts as indication of specific framing, Olausson has developed an analytical tool for frame analysis that is based on critical discourse analysis. It focuses on overarching themes and categories, as well as coherence, implicit information, choice of words and quotes and the rhetorical intentions and devices used in the text. For the purposes of this chapter, Olausson's analytical tool has been adapted and somewhat simplified, focusing on the following key questions:

- *Which themes and topics are granted prominence in the text as a whole, with special attention paid to introductory texts and summaries?*
- *In what ways are claims made about cause-and-effect relationships (attention to the question of attribution of climate change)?*
- *Implicit information. What information is implicit, implied, taken for granted, and dependent on a certain worldview?*
- *Choice of words. Which words are chosen in preference to others?*

In the analysis, particular attention was placed on the treatment of issues related to 'variability' and 'change' in the Arctic and on the specific question of how the issue of attribution was addressed, that is, the links between observed changes in the Arctic and their underlying cause with a focus on the role of human influence.

In order to trace the scientific framing, we have selected a number of key texts that are likely to mirror the scientific consensus at different points in time, such as scientific assessments and science plans that are the result of wide community consultation and/or review. They are the Arctic Climate Impact Assessment (ACIA, 2005) and its popular science summary (ACIA, 2004); the scientific reports from the IPCC, especially the chapters about attribution of climate change (Hegerl et al., 2001; Hegerl et al., 2007) and chapters about polar regions (Anisimov et al., 2001; Anisimov et al., 2007), reports from the Second International Conference on Arctic Research Planning (Bowden et al., 2006), the scientific program for the International Study of Arctic Change (Murray et al., 2010); and the scientific report and executive summary from a study of the Arctic cryosphere: Snow, Water, Ice and Permafrost in the Arctic (SWIPA for short) (AMAP, 2011a, 2011b) with focus on the chapter dealing with sea ice. The context for each of these is described in more detail in the presentation of results. These texts of course represent only a small part of the scientific literature, but nevertheless capture major international scientific syntheses that have been published about Arctic climate change.

These texts represent the scientific framing both before and after the 2007 Arctic sea-ice minimum, as portrayed in texts produced in the science–policy interface and aimed at a broader audience than the closest scientific peers. The year of 2007 is significant in several ways in relation to popular understanding of climate change. In addition to the sea-ice images bringing the Arctic Ocean to public attention, the IPCC released its fourth assessment in the fall of 2007. This reached closure on much of the debate about the human influence on global climate change, which had previously played a prominent role in the political discussion surrounding negotiations in the UNFCCC. In 2007 the public also met the scientific evidence in the popular science format of Al Gore's movie, *An Inconvenient Truth*, and it was the year in which Al Gore and the IPCC shared the Nobel Peace Prize. Media coverage of climate change reached an all-time high (Boykoff and Rajan, 2007).

Framings of Arctic climate change up until 2007

IPCC's third assessment (2001)

IPCC's third assessment, presented in 2001, followed up on IPCC's 1995 conclusions about human influence on global climate change with a statement that 'most of the observed warming over the last 50 years is *likely* to have been due to the increase in greenhouse gas concentrations' (Houghton et al., 2001, p. 61). With today's attention to the impacts of climate change in the Arctic and the notion of the Arctic as a bellwether for global climate change, it is tempting to assume that observed changes in the Arctic played a major role in the scientific discussion about whether climate change was indeed the result of human actions.

However, this assumption is not supported by the analysis of the scientific text in the IPCC reports, where the issue of attribution was prominent long before changes in the Arctic started to receive major attention by the IPCC authors. The attribution issue was much more closely linked to the results of model studies.

Although several statements point to the consistency between observations of change in the Arctic and conclusions based on model studies, the chapter on Polar regions from IPCC's Working Group II also notes that 'It is not yet clear whether changes in sea ice of the past few decades are linked to a natural cycle in climate variability or have resulted explicitly from global warming' (Anisimov et al., 2001, pp. 697, 803).

The framing of the Arctic as a bellwether of global climate change is visible mainly in terms of the Arctic as an extremely vulnerable region, something that had been noted already in IPCC's 1997 Special Report on Regional Impacts of Climate Change. The 2001 report notes that 'the Arctic is extremely vulnerable to projected climate change' and that 'major physical, ecological, sociological, and economic impacts' are expected. The main reason that the report identified was the variety of positive feedback mechanisms that make the Arctic likely to 'respond rapidly and more severely than any other area on Earth, with consequent effects on sea ice, permafrost, and hydrology' (Anisimov et al., 2001, p. 807). The authors reiterate that the Special Report on Regional Impacts (RICC) had 'noted that substantial loss of sea ice in the Arctic Ocean would have major implications for trade and defense' (citing Everett and Fitzharris, 1998).

The Arctic Climate Impact Assessment (2005)

The Arctic Climate Impact Assessment (ACIA) conducted under the auspices of the Arctic Council 2000–2004 has been hailed as a major tide turner regarding attention to Arctic climate change. As described in detail by Nilsson (2007), the ACIA played an important role in placing the Arctic at the center of the global climate change discussion, not least in the media, by framing the issue of climate change as something happening in the here and now rather than as a result of model calculations and concern for the distant future.

Based on the fact that the ACIA played such a major role in portraying the Arctic as a bellwether for global climate change, one might expect that the issue of attribution played an important role in the scientific report from the assessment. This was not the case. There was a lively discussion during the ACIA process about the possibility of making a statement about attribution of Arctic climate change but the view among several lead authors was that this was difficult because of the large natural climate variability in the Arctic (Nilsson, 2003, 2007; ACIA, 2003). The dominant framing in discussing observations of climate change in the past and present in the scientific report instead focused on the variability of the Arctic climate, rather than on the role of greenhouse gases (GHG), as exemplified by the following statement:

> The question is whether there is definitive evidence of an anthropogenic signal in the Arctic. This would require a direct attribution study of the Arctic, which has not yet been done. ... In climate model simulations, the arctic signal resulting from GHG-induced warming is large but the variability (noise) is also large. (McBean, 2005, p. 38)

However, ACIA Chapter 18, 'Summary and Synthesis of the ACIA', uses wording similar to that of the IPCC: 'there is new and strong evidence that in the Arctic much of the observed warming over this period is also due to human activities' (Weller, 2005, p. 991). A similar emphasis on change caused by human action is also apparent in the ACIA overview document, which is a popular science summary of the scientific report: 'Examining the record of past climate conditions indicates that the amount, speed, and pattern of warming experienced in recent decades are indeed unusual and are characteristic of the human-induced increase in greenhouse gases' (ACIA, 2004, p. 23).

It is also in the overview document rather than in the scientific text that the idea of the Arctic as a bellwether for global climate change becomes apparent. It is explicitly stated in the following passage:

> Just as miners once had canaries to warn of rising concentrations of noxious gases, researchers working on climate change rely on arctic sea ice as an early warning system. ... In recent years, the rate of retreat has accelerated, indicating that the canary is in trouble. (ACIA, 2004, p. 24)

This image connected directly to a discourse of the urgency of climate-policy action (Nilsson, 2007). In fact, it appears as if two parallel discourses were becoming established around this time. One of them, focused on change, appears to be directly related to the political sphere and a need to convince policymakers that climate change is indeed happening. In this discourse, the variability of the Arctic climate was not visible as a relevant concern. Instead, it relied on the IPCC and projects the global conclusion onto the Arctic. The other was a scientific discourse that still very much focused on the variability of the Arctic climate. It emphasized that the noise of the system made it very difficult to detect definitive signals showing that the observed changes in the Arctic were indeed caused by emissions of greenhouse gases, that is, attributable to anthropogenic causes. However, it did emphasize that observed changes were consistent with expectations based on climate models.

Research planning: ICARP 2006

The portrayal of the Arctic as a bellwether was further bolstered in a process that took place directly following the ACIA, the Second International Conference on Arctic Research Planning (ICARP-II), which was held in Copenhagen in the fall of 2005 when discussions about the International Polar Year 2007–2008 had just started. The ICARP-II report (Bowden et al., 2006) provides a window into how Arctic climate change was framed in the context of formulating the future direction for Arctic research. The issue of attribution of climate change is not visible in the ICARP-II synthesis report, while a reiteration of the bellwether image suggests that 'change' rather than 'variability' was the dominant framing, as illustrated by the following quote: 'The Arctic has become the bellwether of the Earth's changing climate and other environmental changes' (Bowden et al., 2006, p. 17).

In the summaries of some of the science plans, which may be more representative of scientific discourse than the synthesis, the issue of 'variability' is more visible. Examples include suggestions for research aimed at getting a better understanding of the long-term history of the Arctic Ocean and its gateways. Variability is also an issue in discussions about climate modeling, for example in relation to gaining a better understanding of the feedbacks that underlie the variations in climate. In general, however, the ICARP synthesis publication is indicative of a shift in focus towards change. The issue of variability was still of scientific interest but more for understanding the dynamics of the Arctic climate in the past than in relation to policy processes. Variability does not appear as a counter-discourse to the focus on change and is better described as a parallel track, with its home within the scientific community and with few, if any, immediate implications for climate policy.

IPCC's fourth assessment 2007

IPCC's fourth assessment included strong statements about the link between global climate change and human activities and pretty much concluded the scientific discussion regarding attribution at the global scale. Language about the Arctic was much more careful, as in the following example:

> Anthropogenic forcing has likely contributed to recent decreases in arctic sea ice extent and to glacier retreat. The observed decrease in global snow cover extent and the widespread retreat of glaciers are consistent with warming, and there is evidence that this melting has likely contributed to sea level rise. (Hegerl, 2007, p. 9)

The detection of anthropogenic warming at the continental or subcontinental scale (as opposed to global) was a new issue in the 2007 report and the authors concluded that 'Anthropogenic influence has been detected in every continent except Antarctica' (where lack of observations poses a problem). The Arctic is not a continent and it was thus neither included nor excluded in the general conclusion.

The body of the chapter about attribution highlights the issue of change in the Arctic climate in several passages within an overall framing focusing on variability. Attention to long-term 'change' caused by human action is mainly visible as careful statements about the inconsistency between observed changes and explanations that do not take greenhouse gases into account. One example is a discussion about the observed changes in sea ice with the conclusion that 'observed decreases in arctic sea ice extent have been shown to be inconsistent with simulated internal variability, and consistent with the simulated response to human influence' (Hegerl et al., 2007, p. 718).

In Working Group II's Chapter 15, 'Polar regions (Arctic and Antarctic)', the introduction featured a short discussion about attribution of Arctic climate change citing observation from the ACIA and more specifically a study that came out a year later (Serreze and Francis, 2006). The IPCC authors concluded that 'a substantial proportion of the recent variability is circulation driven, and that the Arctic is in the early stages of a manifestation of a human-induced greenhouse signature' (Anisimov et al., 2007, p. 656). However, the scientifically verified signal of human-caused climate change in the Arctic was still elusive as late as in the preparation of IPCC's 2007 report, a major reason being the high natural variability in the Arctic climate.

Framing of Arctic Climate Change after 2007

The year 2007 was significant for climate science and policy discussions in a number of ways. In addition to the attention caused by the 2007

Arctic Ocean sea-ice minimum, the 2007 IPCC assessment laid the discussion of attribution of climate change at the global scale pretty much to rest. In the Arctic, political attention turned increasingly towards issues of sovereignty, including the infamous planting of a Russian flag on the seafloor of the North Pole and discussion about the potential for new shipping routes. Scientists interested in the polar regions were in the midst of implementing the International Polar Year, including a large number of field studies in the Arctic Ocean. There were also efforts from both the United States and the European Union directed towards a better understanding of the Arctic sea ice, the so-called SEARCH for DAMOCLES program (combining the US SEARCH with the EU DAMOCLES). In this program, the surprise of the sea-ice minimum led to a decision to pool resources to prepare a sea-ice 'outlook' with monthly reports based on a synthesis of observations from different experts (Calder, 2011). Their reports became a convenient open resource for anyone interested in the sea ice, and their data and images have often been used in media presentations.

One question is whether the many direct observations of change in the Arctic Ocean have led to a shift in emphasis from 'variability' to 'change' in the scientific discourse as well. In order to attempt an answer, the analysis of two different syntheses of Arctic climate change science that were published after 2007 is presented below.

International Study of Arctic Change (ISAC)

In 2010, the International Study of Arctic Change (ISAC) published its science program. ISAC is an endeavor that was initiated in 2003 by International Arctic Science Committee (IASC) and the Arctic Ocean Sciences Board (AOSB). In 2005 it presented an initial overview of its priorities, which after review and consultations within the scientific community was finalized in 2010. The purpose of the plan was to 'outline an open-ended international research program and a framework for comprehensive study of arctic environmental change in all its dimensions' (Murray et al., 2010, II). In contrast to ICARP-II, attribution figures prominently in the ISAC Science Plan and is presented in relation to ISAC's intention to build capacity for understanding and predicting Arctic change. However, unlike the earlier discussions of attribution (for example surrounding statements in IPCC's assessment), the issue of attribution was not primarily about political responsibility for mitigation. That human-induced climate change was happening appears as a fact in the introductory text and the issue was more about the extent to which the change is about natural variability. The question of attribution was explicit in the detailed science questions, for example in the following quote:

> The large changes observed in the Arctic environment over the past decades pose a unique challenge to attribution. The inarguable increases of greenhouse gas concentrations and aerosol loadings, globally as well as in the Arctic, are consistent with a warming which may be amplified locally or regionally by feedbacks within the Arctic System.... While anthropogenic forcing does contribute to arctic change there are still unanswered questions concerning its relative importance, especially on the regional and local scales. The complex spatial changes of Arctic temperatures during winter imply that answers concerning attribution will vary by location.... (Murray et al., 2010, p. 26)

In summary, while the issue of attribution was no longer politically as important as it was before 2007, the scientific interest had not diminished by 2010. It might even be relevant to ask whether the political consensus about the need to address the causes of climate change made it easier to re-emphasize attribution. The question is difficult to answer conclusively without further information about how the writers of the reports have reasoned. However, looking at the most recent scientific synthesis of the impacts of climate change in the Arctic at least sheds some further light on how the issue of attribution is treated after 2007.

Snow, Water, Ice and Permafrost in the Arctic (SWIPA)

In 2011 the Arctic Council presented an assessment of changes in the Arctic cryospheres: Snow, Water, Ice and Permafrost in the Arctic, or SWIPA for short. This assessment was not aimed at addressing the questions of attribution of observed changes in the Arctic but took global climate change as a starting point for the analysis, citing IPCC 2007 and the ACIA (AMAP, 2011a, pp. 1–5)

The logic of laying the controversial issues of detection and attribution to rest was further supported with reference to surprises in terms of rapid changes in the cryosphere provided by nature itself, along with the increased capacity of technologies to detect relevant signals in the noise, with the conclusion that "Scientists are now convinced that the emerging picture of a rapidly melting Arctic environment is not a coincidence, but a real and significant change in the climate system of the Earth (ibid.). Moreover, the bellwether image was reiterated in SWIPA's scientific report, here with specific reference to the transformation of the ice cover:

> The Arctic is a bellwether of change in global climate primarily because of the sensitivity of the thin sea-ice cover – which insulates the comparatively warm ocean from the much colder atmosphere – to changes in air temperature and winds. This sensitivity is enhanced by climate feedbacks associated with a changing cryosphere. Dramatic changes are occurring across the Arctic; one of the most visible indicators of this change is the transformation in the sea-ice cover and the associated impacts on climate, ecosystems, and human societies. (AMAP, 2011a, pp. 9–2)

With the issue of attribution placed to the side by the report authors, it might be tempting to draw the conclusion that the SWIPA represents a complete shift from 'variability' to 'change' in the scientific framing of Arctic climate change. However, similar to the parallel emphasis on variability in the ISAC science plan, the SWIPA report also highlights that 'variability' still has an important place in the scientific literature – not least in relation to Arctic sea ice (for example AMAP, 2011a, Chapter 9, p. 3).

The issue of variability is also in focus in discussions about how models can be used for understanding the behavior of sea ice, including the sea-ice minimum of 2007, and the text also indicates that understanding the role of variability is important for understanding what the future holds in store:

> The amount of year-to-year climate variability may be key to the timing of a transition to a seasonally sea ice-free environment.... Higher variability, similar to that of the past two decades, increases the likelihood of summer ice-free conditions by 2050, while lower variability (similar to conditions between the late 1940s and the late 1980s) results in a reduced chance of ice-free conditions by 2050. (AMAP, 2011a, Chapter 9, p. 18)

The SWIPA authors also conclude that a number of different factors were important for the 2007 sea-ice minimum, including both long-term gradual changes in the quality of the ice pack and variability in wind pattern. The framing that emerges is one that places the focus on the fact that the cumulative effects of warming had created a thinner and younger pack ice that made it much more sensitive to perturbation in temperature and wind pattern (AMAP, 2011a, pp. 9–32).

In summary, the SWIPA report illustrates that the issue of variability is still central in the scientific discourse. The text also indicates that the science community views the understanding of variability as important for understanding the future of the Arctic sea ice.

Summary and conclusions from frame analysis

The analysis of scientific assessment, summaries for policymakers and science plans shows that the image of the Arctic as bellwether for climate change appears prominently in 2004 to 2006, especially in writings that are aimed at a wider audience. By contrast, scientific literature is cautious in making explicit assumption that observed changes in the Arctic are the result of anthropogenic climate change as late as 2007. A shift in the scientific literature is visible in writing from 2010 and 2011, where the importance of human driving forces is taken for granted. However, the issue of 'variability' remains important in the scientific literature and figures prominently in scientific efforts to understand what is happening in the Arctic and what it means for the future.

Discussion

Tracing the major framing in the scientific discourse on Arctic warming over the years before and after the 2007 Arctic sea-ice minimum reveals a shift from a very cautious attitude towards attributing Arctic climate change to human causes to a situation in which this is taken for granted as a major driver behind the observed changes. At one level, this shift could be explained by better observations and better technologies, including more powerful climate models and statistical methods that are used for the studies on attribution. Together with the accelerated changes in the environment, they have made it possible to more unequivocally attribute Arctic climate change, including changes in the sea ice, to emissions of greenhouse gases.

However, the frame analysis also reveals a scientific discourse that highlights the variability of Arctic climate change much more strongly than is apparent from media and popular science representations of the Arctic. At a time when the attribution of climate change to human causes was still controversial among political decision-makers, that is, before 2007, this could possibly be the result of a cautious style typical of scientific writing. This might also have been augmented by an extra sensitivity for the need to present only very solid conclusions, that is, conclusions that are beyond possible scientific criticism. Interviews and observations made during the ACIA process indicate that such scientific credibility and extra care was important to the authors. Similar concerns in the climate science community have previously been documented in a study of the use of flux adjustment in climate models (Shackley et al., 1999). The collective writing and review processes of scientific assessments and the science plans would also be likely to emphasize caution in drawing conclusions.

The question is if scientific caution is sufficient as an explanation for the emphasis on variability. If so, one would expect the scientific discourse to place less and less emphasis on 'variability' as the evidence to support attribution to human causes gets stronger. This is not what the empirical material indicates, but quite the opposite: the emphasis on variability remains important and the issue is re-emphasized. Even the specific question of attribution is prominent in material that was released several years after the sea-ice minimum.

This observation leads to two important questions. One is whether the need for clear messages to policymakers before 2007 may have led to a situation where the authors of summaries for policymakers may have underrepresented the relative importance of variability, as it was highlighted in the scientific literature, because it was not seen as policy-relevant in relation to discussion within the UNFCCC climate negotiations. The translations of scientific reasoning into messages to policymakers and the public are by necessity simplifications, and a focus on variability might have been seen as confusing in relation to the more important overall message that the Arctic climate was changing rapidly and the most likely reason was human-caused global climate change. The empirical material from the study of the ACIA process indicates a strong focus on clear and policy-relevant communication. However, no specific questions were asked about the treatment of variability and it is therefore difficult to know why certain choices were made in writing the synthesis and popular science summary.

The second question is whether the noise in which scientists try to find signals of human-caused change may be much more important than is apparent from media impressions of Arctic climate change. This noise, or 'variability', may provide many surprises for anyone who has an image that the Arctic change will continue along a smooth curve. It is not uncommon to hear that controversies in science can hinder scientific progress because the issues at hand are structured in binary terms – anthropogenic climate change, yes or no (Shackley et al., 1999). However, while the binary logic may be representative for media accounts, the frame analysis of scientific assessments and science plans indicates that the scientific community working with Arctic climate change has in fact been able to maintain its emphasis on 'variability' despite a societal context in which it has been important to emphasize the role of human activities.

Concluding reflection

With the scientific controversy surrounding attribution of global climate change laid to rest, the issues of climate variability and attribution of climate change are not very prominent in the popular science accounts of climate change. They are perhaps seen as lacking political significance. However, the consensus that anthropogenic emissions of greenhouse gases matter to the global climate has not led to political breakthrough for mitigation efforts, and accelerating climate change appears inevitable, at least in the short term. This has already led to increasing attention to other concerns, including impacts of climate-related events, vulnerabilities to future climate change, and the increasing need for adaptation. These concerns require types of knowledge other than understanding the role of greenhouse gases for global or Arctic climate change. It includes knowledge about many of the issues that are linked to 'variability', such as dealing with weather extremes and being prepared for surprises. Judging by the most recent major assessment (AMAP 2011a, 2011b), variability remains a key feature of the Arctic climate and its various components, including sea ice. We still lack much basic understanding about the dynamics of Arctic sea ice. Neither warming nor the sea-ice decline is likely to follow a smooth curve. As circulation patterns in the atmosphere and oceans change, surprises are likely. Maybe the time is ripe to highlight the importance of climate variability also in the media and other popular science accounts.

Note

1. These are specific patterns of atmospheric circulation variations, characterized by corresponding pressure distributions, varying in intensity and polarity with time. The Arctic Oscillation is the dominant pattern in Arctic and subpolar areas. The Arctic dipole anomaly is a pattern that sometimes replaces the Arctic Oscillation, letting southern winds into the Arctic.

References

ACIA (2003) 'Summary Report on the Tenth Assessment Steering Committee (ASC) Meeting, 15–16 October 2003, London, U.K.', Arctic Climate Impact

Assessment (http://www.acia.uaf.edu/pages/background.html#ASC%20 Reports), date accessed 31 January 2012.
—— (2004) *Impacts of a Warming Arctic: Arctic Climate Impact Assessment* (Cambridge: Cambridge University Press).
—— (2005) *Arctic Climate Impact Assessment* (Cambridge: Cambridge University Press).
AMAP (2011a) *Snow, Water, Ice and Permafrost in the Arctic (SWIPA) 2011: Climate Change and the Cryosphere* (Oslo: Arctic Monitoring and Assessment Programme) (http://amap.no/swipa/), date accessed 31 January 2012.
—— (2011b) *Snow, Water, Ice and Permafrost in the Arctic (SWIPA) 2011 – Executive Summary* (Oslo: Arctic Monitoring and Assessment Programme).
Anisimov, O., et al. (2001) 'Polar Regions (Arctic and Antarctic)', in J. J. McCarthy et al. (eds), *Climate Change 2001: Impacts, Adaptation, and Vulnerability. Contribution of Working Group II to the Third Assessment Report of the Intergovernmental Panel on Climate Change* (Cambridge and New York: Cambridge University Press), ch. 16, pp. 801–42.
—— (2007) 'Polar regions (Arctic and Antarctic)', in M. L. Parry et al. (eds), *Climate Change 2007: Working Group II: Impacts, Adaptation and Vulnerability. Contribution of Working Group II to the Fourth Assessment Report of the Intergovernmental Panel on Climate Change* (Cambridge: Cambridge University Press), ch. 15, pp. 653–85.
Bengtsson, L., V. A. Semenov and O. M. Johannessen (2004) 'The Early Twentieth-century Warming in the Arctic – A Possible Mechanism', *Journal of Climate*, 17, 4045–57.
Bowden, S. (ed.) (2006) *Second International Conference on Arctic Research Planning (ICARP II). The Arctic System in a Changing World. Conference Proceeding. Copenhagen Denmark, 1–2 November 2005* (Edmonton: Nisku Printing).
Bowden, S., R. W. Corell, S. Hassol and C. Symon (eds) (2006) *Arctic Research: A Global Responsibility. ICARP II Second International Conference on Arctic Research Planning* (Edmonton: McCallum Printing Group, Inc.) (http://aosb.arcticportal.org/icarp_ii/), date accessed 31 January 2012.
Boykoff, M. T., and S. R. Rajan (2007) 'Signals and Noise', *EMBO reports*, 8:3, 207–11.
Calder, J. (2011) 'The Sea Ice Outlook', in I. Krupnik et al. (eds), *Understanding the Earth's Polar Challenges: International Polar Year 2007–2008. Summary by the IPY Joint Committee* (Edmonton: International Council for Science and World Meteorological Organization), ch. 3.6, pp. 405–10.
Crutzen, P. J., and E. F. Stoermer (2000) 'The "Anthropocene"', *Global Change Newsletter*, 41, 17–18.
Csonka, Y., and P. Schweitzer (2004) 'Societies and Cultures: Change and Persistence', in *Arctic Human Development Report* (Akureyri, Iceland: Stefansson Arctic Institute).
Edwards, P. A. (2010) *A Vast Machine: Computer Models, Climate Data, and the Politics of Global Warming* (Cambridge, MA: MIT Press).
Edwards, P. A. and S. H. Schneider (2001) 'Self Governance and Peer Review in Science for Policy: The Case of the IPCC Second Assessment Report', in C. A. Clark and P. N. Edwards (eds), *Changing the Atmosphere. Expert Knowledge and Environmental Governance* (Cambridge, MA: MIT Press), pp. 219–46.

Everett J. T and B.B. Fitzharris (1998) 'The Arctic and the Antarctic,' in IPCC, The Regional Impacts of Climate Change. An Assessment of Volnerability. pp. 85–104.

Gillett, N. P., et al. (2008) 'Attribution of Polar Warming to Human Influence', *Nature Geoscience*, 1, 750–4.

Hegerl, G. C., et al. (1996) 'Detecting Greenhouse Gas Induced Climate Change with an Optimal Fingerprint Method', *Journal of Climate*, 9, 2281–306.

—— (2001) 'Detection of Climate Change and Attribution of Causes', in J. T. Houghton et al. (eds), *Climate Change 2001: The Scientific Basis. Contribution of Working Group I to the Third Assessment Report of the Intergovernmental Panel on Climate Change* (Cambridge and New York: Cambridge University Press), ch. 12, pp. 695–738.

—— (2007) 'Understanding and Attributing Climate Change', in S. Solomon et al. (eds), *Climate Change 2007: The Physical Science Basis. Contribution of Working Group I to the Fourth Assessment Report of the Intergovernmental Panel on Climate Change* (Cambridge: Cambridge University Press), ch. 9, pp. 663–745.

Houghton, J. T., et al. (2001) *Climate Change 2001: The Scientific Basis. Contribution of Working Group I to the Third Assessment Report of the Intergovernmental Panel on Climate Change* (Cambridge: Cambridge University Press).

IPCC (2001) *Climate Change 2001: Synthesis Report. A Contribution of Working Groups I, II, and III to the Third Assessment Report of the Intergovernmental Panel on Climate Change* (Cambridge: Cambridge University Press).

—— (2007a) *Climate Change 2007: Synthesis Report. Contribution of Working Groups I, II and III to the Fourth Assessment Report of the Intergovernmental Panel on Climate Change* (Geneva: IPCC).

—— (2007b) 'Summary for Policymakers', in S. Solomon, D. Qin et al. (eds), *Fourth Assessment Report of the Intergovernmental Panel on Climate Change* (Cambridge: Cambridge University Press).

Jones, P. D., and M. E. Mann (2004) 'Climate over Past Millennia', *Reviews of Geophysics*, 42, doi:10.1029/2003RG000143.

Kattsov, V. M., and E. Källén (2005) 'Future Climate Change: Modeling and Scenarios for the Arctic', in *Arctic Climate Impact Assessment* (Cambridge: Cambridge University Press), pp. 99–150.

König, T. (2006) *Frame Analysis: A Primer* (http://www.lboro.ac.uk/research/mmethods/resources/links/frames_primer.html), date accessed 3 October 2006.

McBean, G. (2005) 'Arctic Climate: Past and Present', in *Arctic Climate Impact Assessment* (Cambridge: Cambridge University Press, p. 21).

Miller, C. A., and P. N. Edwards (eds) (2001) *Changing the Atmosphere: Expert Knowledge and Environmental Governance* (Cambridge, MA: MIT Press).

Min, S-K., X. Zhang, F. W. Zwiers and T. Agnew (2008) 'Human Influence on Arctic Sea Ice Detectable from Early 1990s Onwards', *Geophysical Research Letters*, 35:L21701, doi:10.1029/2008GL035725.

Mitchell, R. B., W. C. Clark and D. W. Cash (2006) 'Information and Influence', in R. B. Mitchell, W. C. Clark, D. W. Cash and N. M. Dickson (eds), *Global Environmental Assessments: Information and Influence* (Cambridge, MA: MIT Press), pp. 307–38.

Murray, M. S., et al. (2010) *International Study of Arctic Change: Science Plan* (Stockholm: ISAC International Program Office).

Nilsson, A. E. (2003) 'Observation notes: Assessment Steering Committee Meeting', London, 15 October 2003. Unpublished.

—— (2007) *A Changing Arctic Climate: Science and Policy in the Arctic Climate Impact Assessment*, PhD dissertation. Dep. of Water and Environmental Studies, Linköping University (Linköping, Sweden: Linköping University Press) (http://urn.kb.se/resolve?urn=urn:nbn:se:liu:diva-8517).

Olausson, U. (2009) 'Global Warming – Global Responsibility? Media Frames of Collective Action and Scientific Certainty', *Public Understanding of Science*, 18, 421–36.

Overland, J. E., and M. Wang (2005) 'The Arctic Climate Paradox: The Recent Decreases of the Arctic Oscillation', *Geophys. Res. Lett.*, 32.

Serreze, M. C., and J. A. Francis (2006) 'The Arctic Amplification Debate', *Climatic Change*, 76, 241–64.

Shackley, S., J. Risbey, P. Stone and B. Wynne (1999) 'Adjusting to Policy Expectation in Climate Change Modeling', *Climatic Change*, 43, 413–54.

Solomon, S., et al. (eds) (2007) *Climate Change 2007: The Physical Science Basis Contribution of Working Group I to the Fourth Assessment Report of the Intergovernmental Panel on Climate Change* (Cambridge: Cambridge University Press).

Stott, P. A., et al. (2010) 'Detection and Attribution of Climate Change: A Regional Perspective', *Wiley Interdisciplinary Reviews: Climate Change*, 1, 192–211.

Von Storch, H., E. Zorita and F. Gonzalez-Rouco (2009) 'Assessment of Three Temperature Reconstruction Methods in the Virtual Reality of a Climate Simulation', *International Journal of Earth Science*, 98, 67–82.

Weller, G. (2005) 'Summary and Synthesis of the ACIA', in *Arctic Climate Impact Assessment* (Cambridge: Cambridge University Press), pp. 989–1020.

6
A Question of Scale: Local versus Pan-Arctic Impacts from Sea-Ice Change

Henry P. Huntington

Introduction

In contrast to the other contributions in this book, this chapter is about the media coverage that wasn't, largely because of the story that wasn't. The dramatic loss of sea ice in the summer of 2007 was a stunning story, a vivid sign that predictions of change were, if anything, too conservative, an ideal visual image of a profound effect of climate change. Striking maps showed how little sea ice was left, how much more open water there was in September 2007 than ever before. Reporters wrote about the missing ice and pundits pronounced upon the doomed polar bears, the new shipping routes, the Russian flag deposited in a titanium case on the seabed at the North Pole. The world was changing: nothing would be the same.

And yet some things were the same. Inuit still hunted seals. Iñupiat and Yupik still hunted whales. The ice returned in the fall and the ocean was again frozen in winter. For all the media attention to the 2007 summer sea-ice minimum, there appeared to be very few stories about its impact on coastal communities, even though over the years the media has hardly ignored such impact. It is difficult to determine why something did *not* happen, but perhaps the local aspects of the rapid ice retreat were not as straightforward and eye-catching as the larger-scale implications that attracted so much attention.

In this chapter I describe some of the ways that Inuit peoples use sea ice and open water; what specific characteristics (ice type, timing of melting and freezing, and so on) govern those uses; why the 2007 retreat was not particularly spectacular when seen from Arctic coasts; why the larger implications of the retreat are perhaps more critical; and how local and global perspectives compare over the short- and long-term. My focus is primarily on northern Alaska, in part because of my greater familiarity with that part of the Arctic, and in part because it is one of the few places where coastal peoples actively use the Arctic Ocean proper, rather than marginal seas, estuaries, or channels between islands.

In addition to publications of many colleagues cited herein, much of the material in this chapter comes from various research projects which have involved me in formal and informal discussions with Arctic residents, as well as personal experiences on the ice or with friends in the Arctic. Some of this material has not previously appeared in print. The observations and results I draw on are the product of systematic inquiry using various methods such as interviews, participant observation, and workshops (for example, Huntington et al., 2009). To tell the story of 2007, I have selected sources and examples that, in my view, best shed light on the factors that shaped Arctic residents' reactions (or lack of reactions) to the summer ice retreat, that best illustrate why there was no special local angle to the story, and that place the event in a wider context of Arctic change. The result, I hope, is a concise, fair appraisal, though assuredly from one person's point of view.

My first exposure to sea ice occurred when I was five and ice formed on Stony Brook Harbor, New York, where my grandparents lived. I slipped on a floe lying on the beach at low tide, cutting a gash above my left eye. Fortunately, I recovered and braved the ice again after university, when I went to Barrow, Alaska, to count bowhead whales. Interactions with the Iñupiaq whalers opened my eyes to indigenous cultures and knowledge, and led to a career studying human–environment interactions in the Arctic. I have been involved in several studies of sea ice in Alaska and Canada, typically seeking ways to link indigenous observations and experience with scientific understanding (for example, George et al., 2004; Gearheard et al., 2006; Huntington et al., 2010). During this time the Arctic has shifted from the periphery of global awareness closer to the center, as climate change and sea-ice retreat have become increasingly pronounced in the region. The idea that summer sea ice might disappear within our lifetimes is an idea that most of us would have found shocking just a few years ago. But not after the summer of 2007.

Local perspectives, indigenous knowledge

This chapter describes local perspectives on sea ice and its retreat in the summer of 2007. The observations and understanding of indigenous

Arctic residents are made and developed in the context of their system of understanding the world and the role of humans therein. 'Indigenous knowledge' and its variants (traditional knowledge, local knowledge, and so on) are terms that attempt to capture that worldview, but often skim past the depths of indigenous ways of knowing and go straight into the jargon of academic publications. A brief introduction is therefore in order to set the stage for the rest of the chapter and its consideration of the differences between local and global views of the summer of 2007.

Indigenous knowledge is described and defined in various ways, but a common version is that it is accumulated from experience over lifetimes and generations and perpetuated within a culture (for instance, Kawagley, 1995). The use of a single term, however, obscures the multiplicity of forms of knowledge around the world, falsely implying a simple dichotomy between 'scientific knowledge' and 'indigenous knowledge' (Agrawal, 1995). Instead, indigenous knowledge should be understood as a category of ways of knowing, distinct in some ways from the strict scientific approach, but reflecting various cultural values and histories. While some insights from indigenous knowledge can be readily transferred to scientific ways of thought, other observations and explanations reflect values, philosophies, and beliefs that are beyond the scope of scientific inquiry (for instance, Kawagley, 1995; Huntington, 2009).

The use of indigenous knowledge in scientific research is growing, but still evolving (for example, Huntington, 2011). In the Arctic, the topic has received considerable attention through major research programs and assessments. With regard to sea ice, changes in hunting practices, coastal erosion, and the like have received attention in scientific assessments such as the Arctic Climate Impact Assessment (ACIA, 2005), the Sea Ice Knowledge and Use (SIKU) program (Krupnik et al., 2010a), the Snow, Water, Ice, and Permafrost in the Arctic assessment (for example, Hovelsrud et al., 2011), among others. As was the case with environmental contaminants in the 1990s, the desire for global action needs to be balanced against exaggerating or causing local fears and concerns (for example, Downie and Fenge, 2003). The dual message from indigenous leaders and communities includes faith in the local ability to adapt (for example, Alex Whiting, in ACIA, 2005), but also a recognition that changes and impacts are mounting, and local adaptation cannot handle every challenge (for instance, George et al., 2004).

The use of indigenous knowledge in major international assessments such as the ACIA is in part the result of the growing political presence of indigenous peoples and their organizations in bodies such as the Arctic Council (for example, Nilsson, 2007). An institutional recognition of and commitment to the inclusion of indigenous peoples and knowledge also pervaded the recent International Polar Year (Krupnik et al., 2011; Eicken and Lovecraft, 2011). The Arctic is among the leading areas of the world in this regard. In the 2007 report of the Intergovernmental Panel on Climate Change (IPCC), for example, indigenous knowledge was included only for Africa and the Arctic, and even in those cases emphasized societal adaptation rather than observational data (Parry et al., 2007).

Those observational data are important, however, offering a great deal of local detail as well as a depth of interpretation based on the years and generations over which indigenous knowledge has been accumulated, tested, and refined (for example, Berkes, 1999). At the same time, the spatial scale of observations must also be taken into account. Indigenous knowledge is 'local' in the sense of being based on experience within a specific locale or region, rather than an attempt to draw general conclusions that cover an entire ecosystem or species (for example, Huntington et al., 2004). As we review the ways in which Arctic coastal residents interact with sea ice, and the way that the sea-ice retreat of 2007 appeared from the shores of the Arctic Ocean, it is important to keep in mind the fact that indigenous knowledge, experience, and use are based on local activity and local phenomena.

Indigenous use of sea ice and open water

The Inuit people – from the Yupik of southwestern Alaska to the Siberian Yupik of Chukotka, Russia; across northern Alaska and Canada to the Inughuit, Kitaamiut, and Tunumiut of Greenland – are predominantly a coastal people. Their world is defined in winter by sea ice, forming in late fall and lasting until the following summer. In contrast to the common southern view that sea ice is a barrier to shipping and the like, Inuit regard sea ice as a platform for travel and hunting, as habitat for the animals upon which they rely, and as an unforgiving but essential part of the landscape they inhabit. Wesley Aiken, an Iñupiaq elder from Barrow, Alaska, describes sea ice as 'a beautiful garden', a place that provides for his people and rewards those who respect it and use it well (personal communication, 2008).

At its maximum winter extent, sea ice around North America covers all of the Arctic Ocean, much of the Bering Sea, the waters amid the Canadian Arctic Archipelago and Hudson Bay, and Baffin Bay southward to Newfoundland. Of the many communities in this vast area, only a handful are located on the coastline of the Arctic Ocean itself.

Figure 6.1 Map of the Arctic
Source: Hugo Ahlenius, Nordpil, 2013.

Most of these are in northern Alaska, with a few more in northwestern Canada. Otherwise, all the coastal communities of Nunavut, Canada, and Greenland, plus those south of the Bering Strait in Alaska, are on other bodies of water. This distinction is important because year-round sea ice exists only in the Arctic Basin, the Canadian Archipelago, and parts of uninhabited northeast Greenland. The communities of Baffin Island, West Greenland, the Bering Sea, and elsewhere have almost always had only seasonal ice.

The sea-ice minimum of 2007 was most dramatic in the Pacific sector of the Arctic Basin, where ice retreated much farther from the coast than ever recorded. In Baffin Bay or the Bering Sea, by contrast, the lack of summer sea ice was normal. The area of newly open water was far away, separated from residents of those areas by a narrow strait or by many islands and waterways. Hunters on St. Lawrence Island who have been keeping a record of sea-ice observations for years (Krupnik et al., 2010b) made no special mention of any of the events of the summer of 2007 (Igor Krupnik, personal communication, 2012).

Even for coastal communities on the Chukchi and Beaufort seas, facing northward to hundreds of kilometers of open water, the summer view was not all that different. To be sure, sea ice used to come and go with winds and currents all summer. In recent years the sea ice has gone away in summer and not returned until fall, but 2007 was not unusual in that regard. Only from an airplane or a satellite was the extent of the 2007 retreat made clear. But no one lives in airplanes or on satellites.

From the point of view of coastal residents, the most significant changes in sea ice in the past few decades have had to do with timing and quality of ice (for example, Krupnik et al., 2010a). Hunters are used to putting away their boats at a certain time of year, and getting ready to travel and hunt atop the ice instead of in the water. Similarly, in spring or summer, the ice goes away, and boating begins. For Arctic residents, sea ice is a feature of winter, and thus changes in sea ice are a product of warmer winters rather than unusual summers (for example, Krupnik et al., 2010b). To focus on sea ice in summer is to miss the point.

What is notable to Arctic residents is what has changed about winter. In fall, the ice forms later, and its formation follows a different pattern. Multi-year ice floes used to arrive on the Arctic coast of Alaska and drift south through the Bering Strait, acting as nuclei for ice formation and anchors for shorefast ice. This is now less common on the northern coast, and rare south of the Bering Strait (for example, Oozeva et al., 2004). While late ice formation is clearly one outcome of massive ice retreat in summer, it is also a product of insufficient ice formation in winter. The loss of true winter conditions is the big story for Arctic communities.

At the same time, the implications of changes in sea ice are not straightforward. In 2004, a group of researchers and local hunters were discussing sea-ice change in Clyde River, Nunavut. The visiting academic researchers were pressing for a summary of the impacts of later freeze and earlier melt, speculating for example about a potential reduction in seal harvests. Joelie Sanguya, an Inuk hunter and researcher, finally got tired of the questions and replied, 'It's not that simple' (Gearheard, 2006). Instead of a direct correlation (for instance, 'shorter ice season, fewer seals'), he explained that the actual impacts were less direct and less tangible. Hunters experienced a sense of dislocation being out in boats at a time of year when they felt that they should be on the ice. But they were still able to get seals.

Sea ice, the Arctic Ocean, and local people

In Barrow, Alaska, on the other hand, a rapid retreat of sea ice in summer can sharply curtail the hunting season for bearded seals. Spring hunting is carried out primarily on the shorefast ice, the sea ice that attaches to the shore and does not move with the currents (one hopes). Barrow's Iñupiat take seals during winter and early spring, and then begin whaling in April. At this time the main concern has typically been a sudden breaking off of shorefast ice, potentially carrying whaling crews and their equipment out to sea (George et al., 2004). More recently, thinner sea ice has posed a new problem: the ice is not strong enough to hold up large whales as they are being hauled onto the ice for butchering, and whalers must make an extra effort to find suitably thick ice. Still, spring whaling continues without serious disruption.

Summer in some years is a different story. As the shorefast ice breaks up, leaving broken pack ice in the area, hunters are able to launch boats from shore and go in search of bearded seals and walrus. In Barrow, walrus are a common but not dependable part of the seasonal hunting round. Bearded seals, on the other hand, are needed for the skins used to cover the umiaq, the traditional whaling boat used from the shorefast ice in spring. If the ice breaks up and then quickly retreats, the opportunity to hunt bearded seals in the broken ice passes quickly, and whalers may not be able to get the seals they need to make new boat covers.

Thus the speed and timing of sea-ice retreat in early summer is more important for local summer hunting success than the area of minimum extent at the end of summer. When the sea ice has opened and boat travel is possible, summer activities begin. If this happens earlier, then summer activities can start earlier and last longer (for example, Krupnik, 2006). But once the sea ice itself has retreated farther than hunters are willing or able to travel, it does not matter how much farther it goes. It is simply beyond reach, and hunters will have to pursue other animals until the ice returns.

For some animals, the retreat of sea ice appears to have a major effect, which in turn may have impacts along the Chukchi coast. Walrus need to rest from time to time to recover from the exertions of swimming and diving for the clams they eat. When summer sea ice remains over relatively shallow water, as is found in much of the Chukchi Sea, the walrus haul out on ice. In 2007 the ice was too far offshore. Thousands of walrus hauled out on land, most of them near the village of Point Lay, a pattern that has continued (Garlich-Miller et al., 2011). Shore haulouts on the Chukotka coast are well known, but in Alaska walrus haulouts were limited to a few animals now and then. The new pattern means a challenge for Point Lay in keeping human disturbance to a minimum, and a challenge to the walrus to avoid disease, trampling, and depleting local prey.

In addition to the biological impact on walrus, the extent of open water has a major physical effect, and allows the wind to generate larger and larger waves. When sea ice covers the water, even as loose pack ice, the waves remain small. When there is plenty of open water, the waves can continue to build until they hit a coastline. In northern Alaska, coastal erosion from fall storms is an ever greater concern. This is not a new problem (for example, Brunner et al., 2004), but climate change has increased the likelihood that the storms will come before the ice has formed. With no ice to shield the coast, the impact of the storms is much greater. Local residents are well aware of the connections between summer sea-ice retreat and walrus haulout behavior, or threats from fall storms (Igor Krupnik, personal communication, 2012).

One village that has gotten considerable media attention for its erosion woes is Shishmaref, Alaska, on the north side of the Seward Peninsula, in the southern Chukchi Sea. Shishmaref has lost several rows of house sites (most of the houses themselves having been dragged away from the eroding bluff), and faces a bleak future. As summer sea-ice minima get smaller, more of the ocean must re-freeze before coastlines are again protected from waves. The timing of freeze-up in relation to fall storms is as much the issue as the extent of open water. After a certain point more open water does not change the wave patterns a great deal.

Many stories about sea-ice loss, whether focusing on the large-scale changes of sea-ice minima or on local dynamics such as erosion or hunting, emphasize the problems that are caused. As Joelie Sanguya suggests, however, the real story is not always so simple. There can be advantages to changes in sea-ice patterns, too. Bowhead whales are hunted during the fall migration as well as in spring. In Barrow, more open water in fall has tended to leave bowheads more spread out, requiring whalers to go farther offshore and to face rougher seas, because waves can build up higher than before. In Savoonga, Alaska, however, the later freeze-up has created an opportunity. Typically, Yupik whalers here have hunted in spring, but the longer open-water season has allowed them to develop a late fall whaling season, too, such that over 40 percent of whales taken in Savoonga in recent years have been fall whales (Noongwook et al., 2007). This has helped make up for less predictable weather conditions in spring, allowing Savoonga to maintain its overall annual whale harvest.

Savoonga's response to the later freeze-up can be seen as a form of adaptation to the new conditions. Open water in November and December – a formerly rare occurrence – now reliably allows for whaling, creating a new pattern of hunting. In contrast, the thinning of spring sea ice has forced whalers to modify their practices by spending more time looking for suitable ice on which to haul out a whale, but the main pattern of spring whaling in Barrow has not changed. The emergence of new patterns can be seen as an indicator that certain thresholds are being crossed, a trend that will be difficult to reverse. At the same time it is important to note that sea ice continues to change and has not yet reached a new stable pattern, and thus adjustment and adaptation will continue to take place for some time to come.

In local terms, however, the 2007 sea-ice minimum was not notable for new or distinct direct impacts on Arctic coastal communities, other than as a continuation of the kinds of changes that people had already been experiencing for several years. In this sense there was no particularly striking local story to tell, and so it is not surprising that media coverage largely ignored the local aspects of ice loss and focused instead on the pan-Arctic scale and on global implications. But local and global scales are not separate, and the big story of 2007 does matter, if indirectly, to Arctic communities.

Sea ice, the Arctic Ocean, and other people

The big sea-ice stories of 2007, covered in depth in other chapters, include threats to wildlife such as polar bears; new opportunities for shipping; the potential for oil and gas development in remote seas; a land (or seabed) grab; and, of course, sea ice as a visible symbol of global climate change, worse even than the models had projected. These are largely stories about people outside the Arctic, concerned about conservation of bears, looking for new business opportunities, or worrying about the fate of the planet. All of these are important. All of them ultimately shape not only the discussion – in the media and elsewhere – about the future of the Arctic, but also will help shape Arctic communities themselves. In this sense the 2007 sea-ice minimum was a catalyst for raising many topics of vital interest both within and beyond the Arctic.

As summer sea ice recedes, new possibilities arise for accessing Arctic waters and the perceived riches in and beneath them. Previous rushes to exploit Arctic resources, for example the oil and baleen of bowhead whales in the nineteenth century (Bockstoce, 1986), led to extensive economic and cultural change in the region. The persistence of sea ice, however, helped keep remote communities and cultures isolated, keeping both the number of immigrants and the environmental impacts of 'modern' activity comparatively modest. The twentieth century saw increasing contact throughout the Arctic with southern cultures, governments, and businesses, and thus growing acculturation.

Nonetheless, at the start of the twenty-first century, Arctic indigenous cultures retain many traditional activities and characteristics. Hunting and fishing contribute substantially to the daily diet, local languages are still spoken (though many are in danger of being lost), and the relationship with ice and water described in the previous section is still intact, along with the extensive knowledge that sustains that relationship. The loss of sea ice creates the potential for even more rapid cultural and economic change in the near future, as traditional ways must adapt and modern opportunities become more prevalent.

Oil and gas activity is perhaps the most pressing and most visible of these possibilities (AMAP, 2008). While Arctic oil and gas production is hardly new, the prospect of industrial-scale development offshore has yet to be realized in most places. The loss of summer sea ice is perhaps not a decisive factor, both because year-round production will still have to deal with winter ice, and because ice is not necessarily the limiting factor to drilling technology. Nonetheless, the perception remains that less ice means easier access and thus greater likelihood of development. In the case of most Arctic areas, this idea is a double-edged sword. On the one hand, development creates economic opportunity, including jobs and government revenue. On the other hand, the possibility of environmental impacts, especially from an oil spill, poses a major risk to cultures already experiencing rapid social and environmental change.

Shipping, too, offers risks and benefits, but the likelihood of large-scale shipping in the Arctic remains a matter of speculation (AMSA, 2009). Yet there is already discussion of shipping lanes through the Bering Strait, and a 'polar code' for commercial vessels is being discussed by the International Maritime Organization (IMO). Arctic communities are aware that shipping is on the horizon, in one form or another, and visits by tour ships are increasing in many locales.

In this sense 2007 did not really provide a glimpse of the future, as there was no particular increase in economic activity from the brief and unexpected retreat of sea ice that summer. Instead, it forced a discussion about what that future might look like. Whether Arctic peoples were ready for such a discussion or not, the media coverage raised the profile of the Arctic around the world, including within the Arctic. Just as, a few years earlier, climate change became a topic one could not avoid in Arctic communities (a status it retains), economic activity is now at the forefront of local awareness. Each year's sea-ice minimum now generates a few obligatory stories and discussion about when (no longer if) the Arctic will be ice-free, but the stories about oil and gas, about shipping, about fishing, about claims to extended continental shelves – these are the steady drumbeat by which Arctic communities mark the progress of the twenty-first century.

2007, short-term and long-term

By itself, the sea-ice minimum of 2007 did not have a major impact on Arctic communities, even those few communities that face the Arctic Basin where some waters were open for the first time in recorded history (and likely much longer). The longer story is the trend towards less and less ice in summer, likely culminating during some September in this century with an ice-free Arctic. The real impacts will take place over this longer period, as Arctic species adapt or fail to adapt to new conditions, as access to those species changes with shifts in distribution, and as coastal erosion and other processes associated with ice or open water also change. Similarly, oil and gas or shipping will not react suddenly to a single year's ice conditions, but will move according to the longer trends of which 2007 was simply a surprisingly advanced symptom.

In the context of that longer trend, however, the 2007 ice minimum marked a shift in perception, an awareness that indeed the Arctic might really become ice-free within current lifetimes, that a new Arctic is upon us. At a February 2012 meeting I attended in Barrow, Alaska, hunters contrasted the past few years with the early 2000s, implying that the early 2000s were the new baseline against which recent change can be assessed. The meeting itself was about walrus haulouts on the Chukchi coast of Alaska, with 2007 marking the start of the huge aggregations near Point Lay. Here was a visible sign of that new Arctic, no longer off in the future somewhere, but here and now, with the haulout pattern continuing in 2009, 2010, and 2011.

If media coverage did not capture the local aspects of the story, perhaps this was because most of the local aspects were not in themselves particularly surprising or new. A continuation of a known trend is not so captivating as a sudden departure from that trend, as was the case with sea ice in September 2007. In the four years of sea-ice minima since then, 2007 has become the benchmark against which other years are measured to see if in fact they constitute a real 'story' or simply a reversion to the 'normal' decline. In these stories the context of the 'new' Arctic is largely implied, but remains nonetheless essential to Arctic residents whose future now seems so closely linked with the fate of summer sea ice.

Looking to that future, the local stories are likely to remain feature stories rather than news. Local impacts will probably take place over time, with occasional bursts such as the erosion of another row of houses in another village, or the movement of large walrus haulouts to another location. There will still be interest in how indigenous traditions are adapting to new conditions, whether environmental or economic, but these adaptations will likely develop bit by bit, community by community, even individual by individual. By contrast, the summer sea-ice minima can be easily placed on a simple graph, compared with previous years, and relied upon as an annual reminder of the state of global climate change impact and an indicator of destruction or possibility depending upon one's point of view. The Arctic is likely to remain visible in the world's eyes, but not necessarily for the same reasons that its changes will remain so important to those who live there.

At the same time, the local story remains inextricably linked to the global one, and vice versa. The loss of sea ice is inherently interesting to some degree, but even more compelling when considered in light of its impact on people and the things people care about (for example, polar bears). The local story is the human backdrop for the global loss of summer sea ice, and the global loss of sea ice drives further attention to Arctic issues of development, which – one way or the other – are critical to the future of Arctic residents.

Dramatic events at either scale may not move in step. The sudden loss of sea ice may not be accompanied by a dramatic coastal event, and a dramatic coastal event may not happen at the same time as an unexpected change in global sea-ice patterns. Media coverage may thus move from local to global and back, but one story underlies both the features on human impact and the news articles on sea-ice loss. The Arctic is changing rapidly, and the future remains unknown.

References

ACIA (2005) *Arctic Climate Impact Assessment* (Cambridge: Cambridge University Press).

Agrawal, A. (1995) 'Dismantling the Divide between Indigenous and Scientific Knowledge', *Development and Change*, 26:3, 413–39.
AMAP (2008) *Arctic Oil and Gas 2007* (Oslo: Arctic Monitoring and Assessment Program).
AMSA (2009) *Arctic Marine Shipping Assessment 2009 Report* (Akureyri, Iceland: Protection of the Arctic Marine Environment).
Berkes, F. (1999) *Sacred Ecology: Traditional Ecological Knowledge and Resource Management* (Philadelphia: Taylor & Francis).
Bockstoce, J. R. (1986) *Whales, Ice and Men* (Seattle: University of Washington).
Brunner, R. D., et al. (2004) 'An Arctic Disaster and Its Policy Implications', *Arctic*, 57, 336–46.
Downie, D. L., and T. Fenge (eds) (2003) *Northern Lights Against POPs: Combating Toxic Threats in the Arctic* (Montreal: McGill-Queen's University Press).
Eicken, H., and A. L. Lovecraft (eds) (2011) *North by 2020: Perspectives on Alaska's Changing Social-Ecological Systems* (Fairbanks: University of Alaska).
Garlich-Miller, J., W. Neakok and J. Herremann (2011) *Field Report: Walrus Carcass Survey, Point Lay Alaska, September 11–15, 2011* (Anchorage: U.S. Fish and Wildlife Service).
Gearheard, S., et al. (2006) '"It's not that simple": A Collaborative Comparison of Sea Ice Environments, Their Uses, Observed Changes, and Adaptations in Barrow, Alaska, USA, and Clyde River, Nunavut, Canada', *Ambio*, 35, 203–11.
George, J. C., et al. (2004) 'Observations on Shorefast Ice Failures in Arctic Alaska and the Responses of the Inupiat Hunting Community', *Arctic*, 57, 363–74.
Hovelsrud, G. K., B. Poppel, B. van Oort and J. D. Reist (2011) 'Arctic Societies, Cultures, and Peoples in a Changing Cryosphere', *Ambio*, 40, 100–110.
Huntington, H. P. (2009) 'Connections between Arctic Peoples and Their Environment', in UNESCO, *Climate Change and Arctic Sustainable Development: Scientific, Social, Cultural, and Educational Challenges* (Paris: UNESCO).
—— (2011) 'The local perspective', *Nature*, 478, 182–3.
Huntington, H. P., T. Callaghan, S. Fox and I. Krupnik (2004) 'Matching Traditional and Scientific Observations to Detect Environmental Change: A Discussion on Arctic Terrestrial Ecosystems', *Ambio Special Report*, 13, 18–23.
Huntington, H. P., S. Gearheard, M. Druckenmiller and A. Mahoney (2009) 'Community-Based Observation Programs and Indigenous and Local Sea Ice Knowledge', in H. Eicken et al. (eds), *Handbook on Field Techniques in Sea-Ice Research (a Sea-Ice System Services Approach)* (Fairbanks: University of Alaska Press).
Huntington, H. P., S. Gearheard and L. Kielsen Holm (2010) 'The Power of Multiple Perspectives: Behind the Scenes of the Siku-Inuit-Hila Project', in I. Krupnik et al. (eds), *SIKU: Knowing Our Ice* (Dordrecht, Netherlands: Springer).
Kawagley, A. O. (1995) *A Yupiaq Worldview: A Pathway To Ecology and Spirit* (Prospect Heights, IL: Waveland Press).
Krupnik, I. (2006) 'We Have Seen These Warm Weathers Before: Indigenous Observations, Archaeology, and the Modeling of Arctic Climate Change', in J. Arneborg and B. Grønnow (eds) *Dynamics of Northern Societies* (Copenhagen: Publications from the National Museum, Studies in Archaeology and History, vol. 10).
Krupnik, I. et al. (eds) (2010a) *SIKU: Knowing Our Ice* (Dordrecht, Netherlands: Springer).

Krupnik, I. L. Apangalook Sr. and P. Apangalook (2010b) '"It's cold, but not cold enough": Observing Ice and Climate Change in Gambell, Alaska, in IPY 2007–2008 and Beyond', in I. Krupnik et al. (eds) *SIKU: Knowing Our Ice* (Dordrecht, Netherlands: Springer).

Krupnik, I. et al. (2011) *Understanding Earth's Polar Challenges: International Polar Year 2007–2008* (Edmonton, AB: Canadian Circumpolar Institute and the University of the Arctic).

Nilsson, A. E. (2007) *A Changing Arctic Climate. Science and Policy in the Arctic Climate Impact Assessment,* PhD dissertation. Dep. of Water and Environmental Studies, Linköping University (Linköping, Sweden: Linköping University Press) (http://urn.kb.se/resolve?urn=urn:nbn:se:liu:diva-8517).

Noongwook, G., the Native Village of Savoonga, the Native Village of Gambell, H. P. Huntington and J. C. George (2007) 'Traditional Knowledge of the Bowhead Whale (*Balaena mysticetus*) around St. Lawrence Island, Alaska', *Arctic,* 60, 47–54.

Oozeva, C., et al. (2004) *Watching Ice and Weather Our Way* (Washington, DC: Arctic Studies Center, Smithsonian Institution).

Parry, M. L., et al. (eds) (2007) *Contribution of Working Group II to the Fourth Assessment Report of the Intergovernmental Panel on Climate Change, 2007* (Cambridge: Cambridge University Press).

7
Under the Ice: Exploring the Arctic's Energy Resources, 1898–1985

Dag Avango and Per Högselius

Introduction

One of the most striking aspects of media's reporting on the retreating and thinning Arctic sea ice has been the insistence that an increasingly ice-free Arctic Ocean creates vast new opportunities for the world's energy supply. The absence of sea ice in summertime is presumed to offer new possibilities for extracting oil and gas reserves, estimated to contain 25 percent of the not yet discovered deposits of those resources, according to the US geological survey (Gautier et al., 2009; AMAP, 2007). Opinions differ as to the consequences: some argue that it offers a welcome relief in the context of global energy scarcity. Others believe it will contribute to delaying a necessary transition to renewable energy sources, while fueling geopolitical tensions in the Arctic region and beyond. The impression is that – real and imagined – energy resources in the Arctic Ocean have stimulated the overall Arctic discourse, boosting the awareness from the side of both governments and the general public of the changing Arctic and, more generally, of the long-term geophysical processes under debate in the context of climate change.

Conversely, the attention paid to the changing sea ice has made the discussion about prospective Arctic energy resources more vivid. Yet this intensifying interest in the Arctic as a promising energy region is not simply an effect of the climate debate. When the 2007 sea-ice minimum aroused worldwide interest, the hunt for coal, oil and gas – as well as a variety of other natural resources that will not be discussed here – had already been going on for more than a century. In fact, as we argue in this chapter, Arctic energy exploration in the early twenty-first century was in a phase of dynamic expansion that would have continued even if global warming had not occurred. To show this, we set out to explore

Arctic energy activities of the past, with a particular emphasis on the ways in which energy actors have dealt with Arctic sea ice – in both practical and rhetorical terms. We argue that the ice – and the harsh northern environment in general – has not only been regarded as a problem, but also as an opportunity. We believe that today's quest for Arctic oil and gas, in the context of climate change, cannot be properly understood without taking these and other historical experiences into account.

The chapter is divided into three main parts: first, we outline the most important long-term trends in Arctic energy exploration; second, we investigate Arctic energy explorers' encounters with sea ice during the era of coal; and, third, in the age of oil and gas. A concluding section discusses these historical experiences in relation to the present-day Arctic energy debate.

Arctic energy in historical perspective

The idea of the Arctic as a region for extracting energy resources is not new. In the opening decades of the twentieth century, several regions in the Arctic caught the interest of actors from the south because they contained coal, the primary energy resource of that time. Coal was the energy provider that fueled the industrialization of Europe and North America. It was used for heating the steam engines, steam turbines and heating processes of factories and works, for the boilers in the locomotives pulling trains on expanding railway systems and steamships of trading fleets, for heating the apartment buildings of growing cities, as well as household stoves. There was a never-ending market for coal. Therefore new finds of this vital energy resource, even in remote regions such as the Arctic, caught the interest of mining industrialists if profitable extraction was feasible (Avango, 2005a).

In some areas of the Arctic, coal mines were opened to cater for local energy needs, while in other areas companies constructed large-scale coal-mining settlements aimed at exports to the industrial centers in the south. One such area was the archipelago of Spitsbergen (Svalbard in Norwegian), located in the European Arctic. Spitsbergen was subject to an intensive rush for coal in the opening decades of the twentieth century, involving mining companies from several nations in Europe and North America – the United States, Norway, Russia, Sweden, the Netherlands, Germany and Great Britain (Avango et al., 2010).

The history of this industry can be divided into three different phases. During an opening phase starting in 1898, entrepreneurs from Germany, Norway and Britain started prospecting companies that opened up small-sized mining camps, operated only during the summer months. During a second phase, from 1905, these companies sold off their coal fields to actors with enough resources (economic, knowledge, networks) to start up large-scale mining operations. The first ones were British (Advent City, 1905) and American (Longyear City, 1906) and as the world-market prices for coal ran high during World War I, mining companies from several countries in the Western world established large-scale mining operations for year-round production – Sveagruvan (Swedish), Barentsburg (Dutch), Brandal City (Norwegian) and Grumant City (Russian) to name a few. This period can be called a coal rush, but it was short. In the mid- and late-1920s it came to an end because of the worldwide economic crisis. The prices of coal dropped rapidly and as a consequence most of the mining companies went bankrupt and closed down their mines. The third mining phase started in the late 1920s and continues to the present. Companies from only two nations have been involved in mining for energy resources on Spitsbergen during this period – Norway and the Soviet Union/Russia. These companies operated several large-scale coal mines each, some of which are still in production (Hacquebord and Avango, 2009).

Coal energy resources were also available in Alaska, known by indigenous peoples of the region for a long time. From the second half of the eighteenth century, actors from Russia and Britain started to explore them during expeditions in the area. The first coal-mining operation in Alaska started already in 1855 by the Russian-American Company, followed by several small-scale mining operations. Most of these mines were opened in southern and central Alaska, but coal mines were also opened in the Arctic parts of the province by whalers and other skippers who used them for the fuel needs of their ships and whaling operations. However, coal mining on a larger scale did not take place until after the construction of the Alaskan railway in 1917, again in southern and central Alaska. In Arctic Alaska, large-scale coal mining came even later, in the early 1940s after the US Department of Defense had issued orders for opening larger-scale coal mining in that area. These coal mines, located along the coast, only supplied a local market and were short-lived (Armstrong et al., 1978, pp. 130–1; Merrit, 1986).

Greenland was also targeted for its energy resources by actors coming from the south. Again, however, the coal was only used for local needs. The first coal mine was opened in 1780 at Disko Bay, western Greenland, and was operated until 1833. In the mid-1920s a Danish coal-mining firm opened a new coal mine at Disko, which remained in operation until 1972 (Armstrong et al., 1978; Bach and Taagholt, 1976).

A more substantial move for Arctic coal energy resources took place in the Soviet Union starting in the 1930s. Already in the late nineteenth century the Russian government had taken initiatives to prospect for coal in the Russian Northwest and on Novaja Zemlja. These prospecting campaigns did not result in any mining operations, but in the 1910s Russian explorers, backed by private capital and the government, found and claimed coal resources on Spitsbergen. As a result, Russian companies opened coal mines at Green Harbor and the Grumant Valley on Spitsbergen (Avango 2005b; Lajus 2004). In the early 1930s, the Soviet authorities opened large-scale mining operations also on the mainland of Arctic Russia, in Vorkuta, as a part of the establishment of the Gulag labor camp system.

In the postwar period, the primary interest in Arctic energy resources shifted from coal to oil and natural gas, whereby Arctic energy explorers pushed towards the north from four different directions: Alaska, Canada, Russia and Norway.

The Alaskan oil industry had been started on a small scale already in the late nineteenth century, but it was the postwar oil surge and the recognition that the United States, from 1949, had become a net importer of oil that stimulated the search to take off in earnest. There was also considerable interest in Alaskan oil from resource-poor Japan. At first activities were concentrated on Alaska's Pacific coast. In the more challenging environment of the North Slope, facing the Arctic Ocean, first drillings took the form of military ventures, but from 1963 activities expanded on a commercial basis. An enormously rich field – North America's largest, as it would turn out – was found at Prudhoe Bay in 1968. The find stimulated further exploration to quickly expand towards both east and west along the Arctic coast.

Prudhoe Bay oil started flowing in 1977. As a result of the dramatic transfigurations on the global oil scene during the 1970s, and steeply rising oil prices, the interest in Alaskan oil from the side of several actor groups continued to increase. The oil hunt thereby also expanded out onto the continental shelf in the Beaufort Sea. A number of fields believed to have commercial potential were found both on- and offshore, but as of the mid-1980s oil companies felt disappointed by the development, as no other giant field had been found. They had expected more. In 1985 Sohio and Exxon became the first companies to decide on the actual exploitation of an Alaskan offshore field – Endicott – but it was not expected to generate much profit. In the oil industry parlor, Endicott was a 'marginal' field. The purpose of developing it was mainly to prevent further exploration from dying off altogether (Curtis and Huxley, 1985).

The neighboring Canadian oil industry traced its origins to the early twentieth century, when oil was found at Norman Wells on the Mackenzie River. Logistical problems made it impossible to distribute the oil in an economically viable way, although it, similar to its Alaskan counterpart, was used for military purposes during and after World War II. Commercial oil and gas exploration in the Canadian Arctic started in the early 1960s and was concentrated on two fronts: the Arctic islands and, further west, the Mackenzie delta. Disappointingly, the actual oil finds were not at all as rewarding as those made in the neighboring US state, and until 1985 the only oil that actually flowed from the Canadian Arctic was from the historical Norman Wells field. That year production from a field on Cameron Island in the High Arctic commenced on a small scale, enabling crude oil for the first time to be shipped through the Northwest Passage to Montreal for refining (*Oil and Gas Journal* [OGJ], 9 September 1985).

Apart from oil, relatively promising natural gas finds had also been made on the Arctic islands, but it was uncertain if and when this gas would ever find any use. A large number of smaller oil and gas fields were also found, but all of them were too insignificant for their exploitation to be commercially viable. The federal government offered generous support to explorers who were willing to continue the search and, in addition, state-owned companies – notably Panarctic Oils – acquired dozens of small private actors that had failed in their Arctic efforts. Precisely the many failures of onshore explorers also stimulated the Canadians, just like their Alaskan counterparts, to expand out into the Beaufort Sea, starting in 1972–1973 (OGJ, 8 March 1971), an effort that was further promoted and accelerated in response to the first oil crisis and the stagnation in US domestic gas production. By the mid-1980s, oil companies were optimistic about the possibility of commercially developing at least a few offshore fields (OGJ, 6 May 1985).

In Europe the interest in Arctic oil and gas built on the tradition already established in Spitsbergen during the coal era. Norwegian, American and Soviet actors arrived in the archipelago during the 1960s with drilling equipment, though the outcome was disappointing from an economic perspective (Barr, 2001). In parallel, stimulated by the huge natural gas finds in the northern Netherlands in 1959, seismic surveys and drilling expanded from continental Europe into the North Sea. The breakthrough for North Sea oil came in 1966. Both British and Norwegian North Sea oil attained special significance in the face of new turmoil on the global oil scene from the late 1960s. The Norwegian strikes were even referred to as Western Europe's 'own North Slope' (OGJ, 18 May 1970 and

25 May 1970). Following the second oil crisis in 1979, the Norwegian government opted to open up waters further north for explorers (OGJ, 15 October 1979 and 23 April 1984). Large oil finds were made in the Norwegian Sea, and in October 1984 Statoil announced its discovery of the Snøhvit oil and gas field in the Barents Sea (OGJ, 4 March 1985).

In Russia, geologists had pointed at the promising prospects for oil in the country's far north already during Imperial times. Lenin sent a hopeful expedition to the Komi Republic in the early 1920s, but it took until the late 1950s before the development took off in earnest. The breakthrough for Arctic oil and gas in the Soviet Union came in the first half of the 1960s through a series of major discoveries in and around the Ob delta and on the Yamal peninsula. The 'Third Baku', as this region was nicknamed, was enormously much richer in both oil and gas than Alaska and the Canadian Arctic. A consequence of this richness was that the Soviets did not feel the same hurry to expand out onto their continental shelf, which thus remained unexplored for much longer than the Beaufort Sea. When *Izvestiya* published a map in March 1966 showing how rich the Arctic Ocean – and especially its Soviet section – was believed to be in terms of oil and gas, it was at the same time emphasized that the first oil and gas from the Arctic offshore would most probably not start flowing before 20 to 30 years, that is, around 1990.

Serious interest in surveys of the Soviet Union's offshore Arctic areas gained momentum only in the late 1970s as Siberian oil – but not Siberian gas – faced the threat of stagnation following over-exploitation of several oil fields. To get a better idea of the geology below the Barents Sea, the Soviets drilled a hole on Kolguyev Island in 1972 and another on an island in the Franz Josef Land archipelago in 1977. From a geological perspective, the Kara Sea was deemed the most promising, but the Barents Sea was more accessible and most early efforts were thus concentrated there. Actual offshore drilling in the Barents Sea was launched in 1982 (OGJ, 22 February 1982). As of 1985 no oil had yet been found, but petroleum geologists were extremely optimistic about coming Soviet offshore finds in the Barents, not least because the Norwegians had made significant finds in their part of the same sea (OGJ, 11 February 1985).

To sum up this section, the present quest for energy resources in the Arctic is not a new development. Over the last 150 years, deposits of coal, oil and gas in the Arctic parts of North America, Europe and Asia have all been interpreted as energy resources and partly also been utilized as such. The motives have been both economic and political. The present efforts to utilize oil and gas resources is merely the latest stage in a more or less continuous development, in which state actors and energy companies have pushed their operations to areas increasingly remote from main centers of energy use.

Encountering sea ice in the era of coal

Industrial companies interested in extracting energy resources in the Arctic have always had to consider the geographical and environmental circumstances of this region when developing their projects. We argue that they have done so in two different ways: by developing technology and through rhetoric.

The coal-mining industry at Spitsbergen in the first half of the twentieth century provides several instructive examples. The mining companies involved were all eager to make maximum use of the opportunities for profits that had opened after the turn of the century. In addition to the general demand for coal described in the above, states tended to support the idea of utilizing Arctic energy in order to secure the needs of their nations, particularly for the railways which were not only an infrastructural backbone in the industrial economy but also had a military function. This issue was particularly problematic in the Scandinavian countries, which lacked satisfactory domestic coal resources. Dependence on imports was not only a security issue but also an unfavorable competitive situation for industry at a time when the free-trade ideas of the mid-nineteenth century had given way to protectionist policies. In addition, there were strong foreign policy interests fueling the Spitsbergen coal rush. Spitsbergen was regarded as a no man's land, therefore the energy resources were available to all, without any state demands to apply for concession or to pay taxes. However, in the opening decade of the twentieth century, Norway challenged the *terra nullius* status with the aim of achieving sovereignty. The Swedish and Russian governments were strongly opposed to these initiatives, which resulted in an international conflict which was not resolved until the competing nations signed the Spitsbergen Treaty in 1920. The states involved in the conflict supported 'their' mining companies and even took initiatives to start new ones, all for the sake of having a strong national presence on Spitsbergen and thereby also a stronger position in the negotiations (Avango, 2005a; Avango et al., 2010). In other words, there was a general economic and political context in Europe which worked in favor of actors who wanted to utilize the coal resources available at Spitsbergen; profits could be made and national needs for energy and political influence could be catered to.

Therefore the coal-mining companies had strong incentives for finding ways to cope with the Arctic environment. First, they did so by developing technology fit for withstanding the demands of this environment. One of the most serious of these challenges was, no doubt, the sea ice. It covered the bays of the archipelago and it appeared as pack ice in the surrounding seas for parts of the year, in particular the summertime. In general the only sites where the mining companies could mine the coal resources in a technologically and economically viable way were at the inner parts of bays, which were ice-covered from late October until the beginning of July. Therefore there was only limited time available for shipping and all coal produced from October to July had to be stored at the mine. For this reason the mining companies could only establish their coal mines at places where there was enough space for storing the entire mass of coal produced at the mines in the winter months – a significant problem in an archipelago characterized by steep coastlines with only limited space available for building industrial works. In addition, the companies had to construct highly effective loading systems in order to be able to load one year's produce of coal onto ships within the narrow window of three ice-free months. Such systems required not

Figure 7.1 Coal-storage area at the Svea mine, Spitsbergen, in the early 1920s
Source: B Walldén photo archive.

Figure 7.2 Pier and loading facilities at Longyear City, Spitsbergen, early 1910s
Source: Longyear Spitsbergen Collection, MTU archives, Michigan.

only space but also substantial investments (Avango, 2005a; Avango et al., 2008a; Avango et al., 2008b).

The sea ice also created other obstacles. The core of the loading systems for coal was the jetties, often equipped with multiple loading systems such as conveyers, railway and aerial ropeways. These complex and expensive loading systems were vulnerable to the movements of the ice. If the ice carried the jetty away while the mine was isolated in the winter, the company would not be able to export their mountain of coal when the summer arrived – a scenario which would lead to bankruptcy. Therefore the companies had permanent work groups whose primary task was to maintain and re-enforce the jetties (Avango, 2005b; Hartnell, 2009).

When sunlight returned to Spitsbergen during the months of spring, the mining companies would often include the sea ice into the technological systems of their mining operations. In general the edge of the Arctic Ocean sea ice was located just off the western coastline of Spitsbergen, some 30 to 50 kilometers away from the mines. The companies utilized the ice edge for anchoring supply ships in order to offload equipment and food to the mines, facilitate the arrival of new employees, and for allowing personnel who wanted to leave an opportunity to do so (or indeed to expel union activists). The supplies were pulled by dog sleds and later bandwagons across the frozen bays to the mines and along the route the companies built intermediary stations for storage and shelter (Avango, 2005a; Avango, 2008a).

Another challenge was the complete absence of vegetation, which meant that mining companies had to import and store massive amounts of wood from the south. Coal mining required a lot of wood, not only to construct buildings for lodgment, services, storage and recreation but also for the production and transport systems. Of crucial importance were wooden pit props, used in huge numbers for supporting the roof of areas in the mine where the coal had been extracted, and most importantly to secure the roof of the main tunnels. The Arctic climate also meant that nothing could be cultivated and that cattle could only be kept with great difficulty. The resources needed to keep a permanent population of hundreds if not thousands of mine-workers alive (and happy), had to be imported. The companies handled this problem by establishing substantial storage facilities and in cases by running greenhouses and keeping live pigs, hens and cows as a living reserve of fresh meat (Avango, 2005a; Avango et al., 2008b).

Another challenge posed by the Arctic environment was the snow. Although there is not much precipitation at Spitsbergen, the open character of the landscape allows winds to carry snow from vast expanses of landscape and deposit it in depressions, behind ridges and at mountainsides. For this reason, coal-mining companies ran the risk of having their buildings and railway systems buried under vast amounts of snow for much of the year. Most companies handled this problem by placing buildings in elevated places, with gabled walls towards the prevailing wind direction, allowing the snow to blow clear from the buildings. They also opted for aerial ropeway transport systems which could carry the energy resources from the mines to the storage areas unhindered by the snow cover. Companies using railways covered them with wooden or concrete tunnels. From the snowfall followed avalanches which posed a particularly serious threat during the early summer thaw. Therefore mining companies usually placed buildings and production systems in naturally protected places and if necessary erected avalanche barriers (Avango, 2004; Avango et al., 2008b; Avango et al., 2009).

Another way of dealing with the environmental conditions of the Arctic was by rhetoric. It is clear that the prevailing discourse about the Arctic in the early twentieth century worked in favor of the Spitsbergen mining companies. They could profit from a contemporary mythology about polar research and exploration as cutting-edge science and as heroism, when rallying political support, investments or when recruiting employees. On the other hand, the same mythology contained stories about encounters with pack ice, isolation and blizzards and these stories were less useful as tools for enrolling investors. Therefore, while using the tropes of Arctic exploration, the mining entrepreneurs and their supporters also tended to downplay the challenges of the Arctic in order to build support for their projects. This became visible in company prospectuses, in correspondence, in professional journals and in the media.

In their rhetoric, mining companies and their supporters tended to emphasize how accessible Spitsbergen was, despite its location in the Arctic. Typically they would refer to the warming effect of the North Atlantic current which rendered the west part of the archipelago free from sea ice for much of the year. Problems with belts of pack ice in the summertime was recognized but downplayed. The Arctic summer would always allow for at least three months of shipping, making it possible to export the annual production of coal from the mines and provide the mining communities with the necessary supplies. The ice cover in the bays, they claimed, could easily be handled by icebreakers, while late-autumn darkness could be countered with a system of lighthouses (Andersson, 1917, pp. 239–48; Brown, 1915, pp. 19–20; Enström, 1923a, pp. 4, 14–16; Hemming, 1921; Högbom, 1914, pp. 199–203).

The 1,000 kilometers distance between the Spitsbergen west coast and the coal market in north Scandinavian ports such as Narvik was recognized, but was described as short in comparison to the closest source of coal to the south: Scania in southern Sweden (Högbom, 1914, pp. 208–10). Spitsbergen was also close to Britain, 'only fifty hours steaming by a fast cruiser from our shores' (Brown, 1915, p. 19). On maps in their coal-mining prospectuses, the mining companies pictured Spitsbergen as located just off the north coast of northern Norway, with shipping routes connecting it with the railway infrastructures of Scandinavia, creating an image of an easily accessible resource at a minor distance from north European markets (Andersson, 1917, p. 248; Enström, 1923a, p. 4; Enström, 1923b; Johnsson, 1916).

In 1919, Scottish geologist and geographer Henry Cadell, a stern supporter of British coal mining on Spitsbergen and board member of the coal-mining firm The Scottish Spitsbergen Syndicate, took this argumentation even further. He predicted that in the future Spitsbergen would suffer even less from the problems of ice. Based on his observations of retreating glaciers and the disappearance of permanent fjord ice

Figure 7.3 Map on the front page of a Swedish coal-mining prospectus from 1916, presenting the Spitsbergen energy resources as easily accessible, ready to plug into the larger infrasystems of Scandinavia. Similar maps are published by scientists and engineers supporting the Swedish mining of coal resources on Spitsbergen
Source: Johnsson, 1916.

in Alaska, he envisioned: 'that the time may come when Norway will see more of the ground free from the grip of glaciers, and the area of the more habitable parts of Spitsbergen considerably extended' (Cadell, 1920).

In cases, mining companies and their supporters presented the Arctic environmental conditions as great assets. The absence of any vegetation made the geological layers readable in the mountainsides, which made it easy for prospecting engineers to map out the extent and character of the coal seams and to identify fault zones. In addition the permafrost kept groundwater from seeping into the mines, which made it unnecessary to invest in expensive drainage pumping systems. Moreover, the low temperatures would make underground mine work comfortable compared to the hot and sweaty character of coal mining in more southern latitudes. Potential problems of wintering over in permanent darkness and complete isolation was recognized, but presented as easily handled by electric light, good housing and sound spare-time activities such as skiing and reading. Indeed, the Arctic climate did nothing but good to those who dared to encounter it: the climate was known to be sound and hardening, even suitable for locating sanatoriums because of a presumed complete absence of any airborne bacteria (Andersson, 1917, pp. 230–4; Brown, 1915, pp. 18-20; De Geer, 1912, p. 372; Högbom, 1914, pp. 204, 206).

In newspaper media, journalists often combined the two narratives. On one hand, their stories romanticized Spitsbergen as an Arctic wilderness, with calving glaciers, polar bears and a landscape beautiful and frightening at the same time. This part of their stories aligned with the wider genre of polar exploration literature at the time and thereby conveyed some of the glow surrounding polar exploration to the Arctic mining industry. On the other hand, the same media stories emphasized the success of the mining companies to handle the Arctic conditions through state-of-the-art engineering. By adapting the latest innovations in transport and mining technologies, the challenges of the Arctic environment – including the sea ice – could easily be overcome. The same was said to be true for the working conditions in Arctic mining settlements. The frustration over isolation and months of winter darkness was supposedly easy to counter through social engineering – decent housing and food, high salaries, libraries, outdoor activities and choirs (Avango and Houltz, 2012).

This rhetoric often had a nationalist content. Through the media, Swedish mining companies and their supporters described the mining industry at Spitsbergen as a result and continuation of a much longer history of Swedish science and exploration in the Arctic. Swedish scientists had played a prominent role in the arena of scientific research in the European Arctic since the 1850s and much of this research had taken place at Spitsbergen. When the geographer Gerard de Geer summarized these efforts in 1912, he counted 24 expeditions, 376 publications and 60 maps (De Geer, 1912, p. 367). Swedish geographers and geologists had not only mapped the features of the Spitsbergen landscape and the depths of its bays, but also the geology, including the coal. Building on this track record, supporters of Swedish coal mining on Spitsbergen portrayed the Spitsbergen coal-mining industry as a result of the efforts of Swedish polar science, highlighting renowned scientists such as A. E. Nordenskjöld, A. G. Nathorst and Gerard De Geer: 'without exaggeration it can be said that what Spitsbergen really is, is what Swedish researchers have made it to be' (De Geer, 1919; Lundström, 1920; von Holstein, 1920). By statements such as these in the media, the mining industry was presented as being built on a solid platform of scientific knowledge, a claim that harmonized well with widespread ideas on the relationship between scientific research and industrial development at the time.

Moreover, the coal industry was built on Swedish knowledge, greater than that of other nations because of supposedly superior standards in science and engineering and an inherent ability of Swedes to cope with the Arctic environment. This attitude was conveyed in newspaper articles and in literature comparing the Swedish Svea mine to other mining towns on Spitsbergen, in which the Swedish Svea mine stood out as the most solid and productive mining town on the archipelago. If anyone, Swedes were able to successfully colonize and utilize the Arctic, as expressed by royal family member and explorer Prince Wilhelm, in a newspaper aimed at Swedish youth in the early twentieth century:

> Some talented engineers and supervisors direct the work, which is exclusively performed by a Swedish workforce. There are medical doctors at the site, and the wireless telegraph system mediates communication with the outside world. In any weather, in rain and snow, in biting cold and arctic darkness, they work the whole year through. It's a constant struggle for survival and keeping the vile climatic conditions at bay. Honor to the men, who thereby establish the northernmost outpost of the Swedish spirit of enterprise; fame to their deeds, which show that where enduring power and purposeful organization is put to work, a Swede will never need to be ashamed of his countrymen, even if the task is to colonize the North Pole! (Wilhelm, 1919, pp. 35–6)

These narratives of Swedish abilities to utilize the energy resources of the Arctic should be understood as tools for attracting investors to Swedish coal mining on Spitsbergen, and to strengthen the Swedish position in the ongoing negotiations regarding the future legal status of Spitsbergen. Similar objectives lay behind much of the rhetoric of other actors within coal mining on Spitsbergen (see for example Cadell, 1920).

However, rhetoric and technology did not always solve the problem of successfully extracting energy resources in the Arctic. There are many examples of mining companies failing to cope with the environmental conditions at Spitsbergen. Often the sea ice was the source of the problem. Mining companies failed to supply their mining communities with the necessities for the wintertime isolation, because sea ice blocked the ships carrying the supplies. Coal ships were forced to turn back to the European mainland in the middle of the summer after meeting solid pack ice instead of an open sea and ice-free bays (Avango, 2005b; Dole, 1922a, pp. 298, 309, 355–6). One of the most extraordinary cases from the coal boom on Spitsbergen in the early 1900s was the summer of 1915, when pack ice blocked the coastline of Spitsbergen during the entire summer. The ships of the mining companies were unable to reach the mines or were crushed in the ice (Dole, 1922b, pp. 372–5). These ice conditions certainly affected the enthusiasm for Spitsbergen as a provider of energy resources. At the time a US firm, the Arctic Coal Company, was trying to sell its coal mine Longyear City – by far the largest coal mine at Spitsbergen – to a Russian syndicate, which sent an engineer to inspect the facilities. His ship got stuck in the ice and two months after leaving Tromsø, it was finally crushed by the ice and his team was forced to walk across the pack ice off the west coast of Spitsbergen, looking for help. Naturally the Russian investors pulled out of the affair and instead the Arctic Coal Company entered negotiations with the Norwegian investors who eventually bought the mining settlement, which today is the provincial capital of Norwegian Svalbard – Longyearbyen (Dole, 1922b, pp. 412–14). Others, however, were not discouraged. The ships of the Swedish mining company spent two months travelling up and down the edge of the pack ice, unable to reach the site where they were constructing their mine (AB Isfjorden-Belsund, 1915). The company board remained silent about this event however. It was not communicated to the shareholders in the company in the annual report and, despite the severity of this setback, the company leadership proceeded to establish what became one of the most significant mining towns on Spitsbergen – the Svea mine. Thus the mining companies tended to exclude narratives about the ice which were not beneficial to them.

In conclusion, the early twentieth century quest for Arctic energy resources did not take place within the context of a debate on global warming and melting sea ice. However, accessibility was an issue in the debate regarding the potential of Spitsbergen as a mining region. The mining companies handled this problem by developing technologies that in most cases were sufficient to withstand the Arctic environment. Equally important, they managed to convincingly construct a narrative about the Arctic conditions in general and the sea ice in particular, in which they were able to successfully domesticate the Arctic conditions through scientific research and modern industrial technology. The ice was turned into a challenge showing their ability.

Oil, gas and ice

In the postwar era, as we have seen, oil rather than coal was the focus. The Arctic energy challenge now became even more daunting. On land, glaciers and permafrost covered many promising oil regions, and when explorers moved offshore they had to deal, depending on the season, with both landfast ice along the coasts and floating sea ice. In both West and East, however, actors believed technology could be developed that would allow exploitation of both onshore and offshore Arctic energy resources. Offshore activities were regarded as the most difficult, but if there was a perceived need to be patient about the realization of this possibility, it was not because actors hoped that global warming – the possibility of which was not widely discussed before the late 1980s – would help to make conditions less harsh, but because scientific and technological progress needed time.

The first challenge was not so much to enable exploration as such, but to overcome logistical barriers for shipments of oil and gas from onshore wells across Arctic waters to worldwide markets, and to bring in machinery and equipment to offshore drilling and production sites. Prudhoe Bay, Alaska's (and North America's) largest oil field, remained unproductive for nearly a decade in the absence of acceptable solutions regarding distribution of its abundant oil. Two main options were considered: to pump the oil through a pipeline overland to US markets, or develop an Arctic tanker route (OGJ, 8 June 1970). Since sea ice made tanker transport difficult, risky or, as some believed, simply impossible during much of the year, stakeholders eventually opted for the pipeline solution, which for its part became one of the most environmentally controversial projects in American history (Coates, 1993). But the sea route was still used for shipping a variety of heavy machinery and equipment to the drill sites. The Alaskan 'sealift' became an annual, almost festive, event for the oil companies involved. From a winter base near Seattle, a convoy of several dozen large barges, loaded with a hundred thousand tons or so of heavy equipment and even housing modules, embarked every year in June or July on their 6,000 kilometer voyage. Having passed through the Bering Strait, the vessels waited at Point Barrow for the Arctic sea ice to break up (this usually happened in August) before continuing to the Prudhoe Bay area. The sealift was an expensive and risky venture. In 1975, many of the barges were 'forced to turn back because of heavy ice', and others were forced to remain through the winter at Prudhoe Bay. The following year the companies managed to hire the service of US Coast Guard icebreakers to prevent this debacle from being repeated (OGJ, 5 July 1976).

The development in the Soviet Union largely paralleled the American one, although the focus here was more on natural gas than on oil. The Soviets took into operation their first long-distance pipelines from onshore Arctic gas fields already a few years before the Trans-Alaska Pipeline was inaugurated. The huge costs of the system, however, were subject to intense internal debate that was played out not only behind the locked doors of the totalitarian regime, but openly in official media. *Pravda*, for example, published a variety of articles that criticized the Arctic pipelines on both economic and environmental grounds (Slavkina, 2002; Gustafson, 1989).

A possible alternative to an Arctic pipeline system, the critics argued, would be to locate major industrial gas users to the Soviet Arctic, as this would eliminate the need for long-distance transport of gas. Another proposal was to move the gas to Arctic harbors for liquefaction and tanker transport abroad. This was an attractive idea in view of the enormous difficulties, delays and breakdowns that the Soviet Union experienced when trying to sell natural gas to foreign customers by pipeline (Högselius, 2012). It seemed to fit well with ideas discussed at the time concerning Soviet gas exports to the United States. Several US gas companies approached Moscow in the early 1970s regarding the possible establishment of a Soviet–American trade in liquid natural gas (LNG), identifying such a trade as a possible solution to America's looming gas shortage. When leading Soviet newspapers in 1976 proudly reported that the Soviet Union was about to make use of nuclear-powered icebreakers to open up 'a year-round shipping route from Murmansk through the Barents and Kara Seas', the Western oil industry was enthused. However, their analysts suspected that the Pechora and Kara seas were much too shallow for large LNG supertankers to operate safely. Instead, if the United States wanted access to Soviet gas, pipelines would first have to be laid from the Siberian gas fields to a harbor further west, on the Barents coast (OGJ, 29 May 1972 and 22 October 1973).[1]

The vision of Soviet LNG exports faded away as geopolitical tensions increased in the late 1970s. From an internal Soviet perspective, however, the opening up of the northern seaway was still of great importance for the country's oil and gas industry. This was because it facilitated a Siberian sealift, similar to the Alaskan one, of heavy equipment, machinery and housing modules to the main energy sites, located as they were in otherwise highly inaccessible regions. Starting in the second half of the 1970s, pressed by the looming scarcity of oil, the Soviets started transporting massive equipment by barge to Cape Kharasavei on the Yamal peninsula's southwestern coast, and from there by specially designed vehicles on the Kara Sea ice pack to promising drill sites on the peninsula's northern tip, several hundred kilometers away (OGJ, 27 June 1977).

The Canadians, meanwhile, aimed to feed oil and gas from the Mackenzie delta into newly built pipelines that would link up with existing pipeline grids. As for the promising gas finds made on the remote Melville and King Christian islands, however, there was much hesitation about the economic feasibility of a piped infrastructure for bringing the fuel southward, the distance to the main consumption centers measuring about 4,000 kilometers. The Canadians had ample experience in ultra-long pipeline construction, having completed a Trans-Canada pipeline already in the 1950s, but the harsh Arctic environment made a corresponding system in a north-south direction a challenge of totally different dimensions. As in Siberia, pipes, compressor stations and other equipment would have to resist extreme temperatures and winds, but in addition the pipeline would have to be protected from potential sea-ice scour on the critical subsea crossings of several Arctic straits – the longest of which measured 122 kilometers across (between Melville and Victoria islands) (Kaustinen, 1983). Research showed that ice scour could be a problem down to depths of at least 50 meters, which meant that pipelines would have to be buried in deep trenches on the sea bottom for protection (OGJ, 3 January 1983).

Another problem, which the Russians never had to face, was an uncertainty as to whether sufficient gas would actually be available. The main stakeholders – of which Panarctic Oils and TransCanada PipeLines took leading roles – formed the Polar Gas Project consortium in 1972 to investigate the issue in depth (Kaustinen, 1983; OGJ, 2 December 1974). The main problem, in Panarctic's view, was not so much the harsh Arctic climate, physically speaking, but rather the lack of a 'suitable *political* and *economic* climate'. Optimism increased with the unprecedented rise in fuel prices, since this meant that the pipeline might pay for itself even though fairly modest volumes of gas were available. In what followed, additional drilling ventures – both domestic and foreign ones – were attracted to the Canadian Arctic. By the early 1980s this new dynamism had resulted in several new gas finds.

Gas shipments in the form of LNG from the Arctic islands were also discussed as an option. An LNG route, it was argued, would give the Canadians greater freedom and flexibility in their export activities and make them less dependent on the United States as their main customer. Envisaging gas sales to Western Europe, Panarctic in 1982 launched a cooperative venture with Germany's largest gas company, Ruhrgas. The main technical challenge when considering the LNG opportunity was to design icebreaking LNG tankers (OGJ, 3 January 1983). In November 1981, a Canadian interdepartmental committee approved a pilot LNG project, saying it trusted that 'the project can proceed safely if a number of recommendations are followed'. These included a 'year round reporting system on ice conditions' and arrangements for providing navigators with 'adequate warning of collision hazards'. There was opposition to the project, however, from the side of environmental groups. Moreover, Greenland's parliament 'voted unanimously to condemn the project and wanted to take the issue to the UN, worried about potential environmental hazards posed by heavy traffic of LNG tankers along its shoreline' (OGJ, 2 November 1981). In August 1984, then, Canada's National Energy Board officially shelved the Arctic LNG pilot project (OGJ, 20 August 1984), though apparently not so much for environmental reasons but rather because of failure to come to agreement with prospective customers.

In the Spitsbergen archipelago, onshore drillings for oil were made difficult both by lack of harbors and by the presence of landfast sea ice. In addition, the Spitsbergen Treaty stipulated that aircraft landing strips must not be built, thus further hampering logistical operations. Even so, the archipelago attracted explorers. In June 1972, for example, a Belgian–Norwegian drilling expedition arrived on Spitsbergen's Edgeøya, which had been identified as a geologically promising site but where landfast ice precluded the reliance on ships for bringing the drilling rig onshore. Instead, all parts of the rig, along with a complete camp for the oil workers, were lifted from the ship to the drilling site by helicopter – a tedious and expensive method. Nevertheless, in 1973 the venture moved on to the even less accessible Hopen Island (OGJ, 10 September 1973). A total of 12 sites were drilled in Spitsbergen during these years, but the lack of actual strikes then led most companies to abandon the contested archipelago. To the extent that an interest was retained, this was because there was 'a fund of geological data to be gathered that could be valuable in preliminary mapping of the Barents Sea' (OGJ, 4 February 1985; Barr, 2001).

In all the projects described above, oil and gas companies were forced to deal with Arctic sea ice despite the fact that drilling and production took place onshore. When the companies in a second phase moved out into the sea itself in search of oil and gas, their ice encounters were further complicated. Again, however, petroleum geologists and engineers did not see any reason to stand back and wait for climate change to free the Arctic Ocean from its icy roof. Scientific advancements and engineering ingenuity were seen to offer solutions to all ice-related problems.

Regular offshore drilling rigs could not be used, vulnerable as they were to the onslaught of sea ice. Instead, the first drillings in waters with seasonal ice cover made use of specially designed drillships, which were anchored up at the respective drill sites for a few summer months, after which they were moved back to ice-free harbors, only to return the next year. In regions where ice was likely to be encountered also in summer, the ships were designed in such a way that they could quickly be moved to a safe location, should dangerous ice be approaching. Alternatively, the ship was accompanied by one or more support vessels whose task, in case of ice encroachment, was to break up the ice around the drill ship (OGJ, 5 August 1985). But seasonally used drill ships, with limited time on the site every year and expensive voyages back and forth between summer and winter locations, were clearly not optimal from a business point of view. Operators were often frustrated by the unpredictability of the year's work, as early winter-ice buildup often forced them to return home weeks and even months ahead of schedule (for example OGJ, 25 June 1984).[2] A first key challenge for operators was thus to find ways of extending the short drilling season and, ideally, enable operation in Arctic waters on a year-round basis.

A central innovation in this context was the artificial drilling island. Built from sand and gravel, it offered a stable operating platform able to withstand the assault of ice. Such islands could only be built in relatively shallow waters (down to depths of 30 meters or so), but even so they enabled the Arctic oil industry to expand their shelf-based activities considerably. The first drilling islands were built in the Beaufort Sea in the mid-1970s and were constructed from sea sand with the help of powerful dredgers. Given the frozen northern seaways, the latter typically had to be brought in from their regular sites of operation on North America's sandy east coast by way of barge transport through the Panama Canal and Bering Strait (OGJ, 6 October 1975). The Soviet Union subsequently also adopted the technology, its first artificial drilling islands being built in 1982 in the Okhotsk Sea (OGJ, 17 May 1982).

By the early 1980s no offshore Arctic oil or gas had yet started flowing and no commercially viable oil field had been found in the Beaufort Sea. Precisely this failure, however, in combination with record-high oil prices and massive R & D support from the Canadian government (OGJ, 20 August 1984),[3] stimulated further efforts in the field of island construction. Increasingly sophisticated island designs fueled a considerable optimism concerning the prospects for important offshore discoveries. A dedicated research station was built on Hans Island, located in the Kennedy Channel separating Greenland from Ellesemere Island. Described as 'a barren, doughnut-shaped rock ½ mile in diameter', it was deemed suitable for island-construction experiments. Ice researchers used it to 'gather data on how multiyear ice affects large structures and how this information could be applied to island building technology'. Since ownership of Hans Island was subject to dispute between Canada and Denmark, however, the erection of the research station may also be interpreted in a geopolitical context.

The oil industry was reportedly 'confident that artificial islands can be built to handle the stress of ice forces', and technical progress in the field was certainly impressive. Whereas the first islands had been designed from natural material, which formed beaches on the island's edges, more advanced constructions took the form of concrete and steel caisson islands, the first of which was built at a depth of 22 meters in the Canadian Beaufort in 1981. The caissons had steep external walls, allowing large savings of sand and reducing the need for dredging, which was here limited to the construction of a subsea sand berm as a platform onto which the caisson was fixed. Smaller versions were still dependent on support vessels for 'ice management', whereas the larger ones looked like massive fortresses planted into the sea.

Arctic oil and gas engineers could also build on an already established tradition of 'iceberg management' developed primarily by the Canadians in non-Arctic operations off Nova Scotia. In the 1980s several Norwegian technology companies started to take interest in this development, forming joint ventures with more experienced US and Canadian firms aimed at new ways of securing offshore rigs against both small and large icebergs. The Gulf Stream made icebergs very rare in the North and Norwegian seas, but the Norwegians already planned to move further north and wanted to prepare for future ice encounters. The most radical innovation was probably the 'fender platform', which used a sawtoothed configuration able to 'withstand the impact of small icebergs or larger ones that might be encountered once every 100 years' (OGJ, 17 January 1983).

A next step was to develop rigs and platforms able to withstand not only the attack of icebergs, but also to cut their way through thick sea ice on their way to a drill site and, once in place, continue defending

Figure 7.4 Sketch of the 'fender platform', developed in the early 1980s by Norwegian and Canadian companies
Source: OGJ, 17 January 1983.

Figure 7.5 Built to cut through thick sea ice en route to drill sites and to withstand moving pack ice during drilling operations
Source: OGJ, 24 June 1985.

themselves against moving pack ice. The lead was here taken by actors involved in exploiting the Beaufort shelf. Canada's Panarctic conducted research on possible rig designs, focusing primarily on development of a 'drilling rig supported on a towed air cushion vehicle and a dynamically positioned, ice cutting semisubmersible rig'. The unit would 'cut its way through the ice on the way to a drillsite, then cut ice to maintain location while drilling if the ice moved'. As of 1983, laboratory tests and Arctic fieldwork was carried out to determine the feasibility of the project (OGJ, 3 January 1983). Odeco, a US drilling contractor, followed this up by developing a 'massive deepwater Arctic rig' with conical body. Its geometry was seen to pacify the ice more effectively than the normal, cylindrical rig structures used in ice-free waters. The structure weighed some 190,000 tons and was placed directly with its large base onto the sea bottom (OGJ, 24 June 1985).

Overall, engineers and companies were thus confident that ice could be 'managed'. But the harsh Arctic climate and the omnipresence of ice was not always viewed as a problem. *Izvestiya*, discussing the possibility of oil production on the Soviet Union's Arctic coast, argued in 1966 that:

> the Arctic oil industry will have to face many difficulties. But they are not as big as we sometimes think. Producing oil on Arctic shores will even be easier than, say, on the Western Siberian lowland. Permafrost, which often becomes a source of all kinds of angers, will here play a positive role, facilitating drilling and transport. The Northern ocean seaway opens up the way for oil to all parts of the world. (*Izvestiya*, 2 March 1966)

The fact that the Arctic was very sparsely populated was also a major advantage, as projects were rarely troubled by the endless bureaucratic and legal problems encountered on more southerly latitudes in terms of necessary permissions to drill wells and lay pipelines on privately owned land. The local communities that did inhabit the Arctic region often showed themselves worried about the possible negative effects on their environment and culture, but the dimension of the problem was not at all the same as in densely populated provinces. Moreover, the oil industry argued that the harsh climate enhanced environmental security. Dome Petroleum, a leading Canadian oil company, explained in 1983 that 'oil spills are easier to deal with in frozen waters. Oil would be sandwiched between ice layers and would stay there during the winter', giving cleanup operations ample time to plan and carry out their work (OGJ, 3 January 1983). From this perspective, retreating and thinning of the sea ice as a result of global warming was as much a threat as an opportunity for Arctic oil and gas production.

Panarctic, for its part, in the early 1980s developed a new drilling method that it intended to use at sites where water depths ranged to 400 meters, that is, much greater than what was possible with the artificial island method. The new technique made use of the permanent Arctic sea ice and foresaw construction of an 'ice pad drilling base' as much as 6 to 7 meters thick, built 'by flooding and freezing seawater' in combination with 'large blocks of low density urethane in the structure of the ice island to increase the load carrying capacity of the platform and reduce construction time'. The success of the method, however, hinged on the lateral stability of the sea ice: if the ice moved by more than 5 percent of the water depth, the drilling equipment risked being destroyed. Before the method was employed at a site, prospective drillers needed to spend at least two years measuring the movement of the ice with the help of 'automated, computerized stations using overhead satellites for positioning' (OGJ, 3 January 1983). Global warming, to the extent that it loosened more and more sea ice and reduced its stability, tended to reduce the number of sites where the method could be employed.

Concluding discussion

In late 1985, when the first US offshore Arctic field was slated for exploitation and the first crude oil from Canada's Arctic islands had just arrived in Montreal, many industry representatives believed that Arctic oil and gas could expect a brilliant future. North America led the development, and Norway and Russia were eagerly following. Actual finds of oil and gas in the North American Arctic had been quite disappointing, but precisely the many 'dry holes' and 'marginal fields' had stimulated petroleum engineers to develop remarkable technological solutions intended to improve the prospects for success in the harsh northern environment. The first half of the 1980s had seen a burst of activity in this respect, fueled by the second oil crisis. Pushing the Arctic energy frontier was excessively expensive, but high oil prices and technological progress made attempts to explore even the most inaccessible energy regions appear worthwhile. 'The key to commerciality', it was firmly believed, 'will be cost containment through innovation'. Arctic oil and gas was thereby seen to show the way for the industry in a more general sense, as 'the revolution of creativity in the Arctic' would most likely 'spread to other sectors of the petroleum industry' (OGJ, 5 August 1985). What can be seen here is how oil companies and the petroleum engineering community made rhetorical use of the harsh Arctic climate to attract investment, including R & D funding from government sources.

Then, in 1986, oil prices unexpectedly plummeted. The 'reverse oil price shock', as it was sometimes referred to, was devastating for the economics of Arctic energy. For consumers the fall in prices was welcome, but for the oil industry it was a disaster. North American companies, which had led the development, were particularly exposed. Suddenly it became apparent how vulnerable operators in the Beaufort Sea and on the Arctic islands were to turmoil on world markets. Companies opted to leave their exploratory sites, plugging the holes that had been drilled at vast cost and abandoning them – at least for the time being. Some actors moved their operations to the Norwegian and Barents seas, which during the second half of the 1980s emerged as the Arctic's new drilling hot spots. The absence of ice here made it feasible to sustain exploration in spite of the depressed market.

It would take until the late 1990s before fuel prices turned upward again. When they did, the world's energy companies felt motivated to invest, once again, in research, development and exploration in offshore areas above the summer ice limit. However, the global political and economic environment was no longer the same. The world economy had become more globalized and dynamic; neo-liberalism had established itself as a political dogma almost everywhere; the Soviet communist bloc had collapsed; China was emerging as a new world power; and the information age was just experiencing its breakthrough. In addition, there was a new surge in environmentalism and a growing concern with global warming.

Analyzing the dynamics of this more recent period is beyond the scope of this chapter, but it might still be commented upon from a historical point of view. Currently, the reported retreating and thinning of Arctic sea ice is at the very heart of the Arctic energy debate, particularly if the general news media are taken as the main point of reference. This is a new theme, but the attempts to exploit the Arctic's energy resources are anything but new. As we have shown, they did by no means start in the age of the climate debate. There is a striking continuity, from the late nineteenth century to the present, in Arctic energy exploration, and the current development is merely the latest addition to an impressive historical trajectory. During this longer period, which started in an era when coal was the world's most strategic fuel, actors repeatedly set out to make use of Arctic fuel without expecting global warming to assist them in making it more accessible. It thus appears plausible that the current quest for Arctic oil and gas would have come about even in the absence of climate change. Technological development and high fuel prices would have sufficed to sustain the race.

In the period analyzed, actors dealt with Arctic sea ice in a practical, technological sense, and simultaneously used it for rhetorical purposes. Depending on the situation, it was argued that ice offered unique opportunities, constituted a problem that needed to be dealt with and funded, or was no problem at all. In the era of coal, it was important for energy companies to assure investors that sea ice did not pose a risk to operations in Spitsbergen and elsewhere, while at the same time the harsh Arctic environment was deliberately invoked to produce an image of the industry as cutting-edge engineering, pushing the frontier of economic development and modernity into the Arctic. This image of the iced Arctic as a region that needed to be conquered and tamed by energy explorers lived on into the age of oil and gas. The Arctic was framed discursively as an energy frontier and an important laboratory in which new technologies could be pioneered to be used later on in other parts of the world as well. But ice could also be a resource in its own right, transformed – materially and conceptually – into artifacts that formed integral parts of drilling, production and transport equipment. Moreover, Arctic sea ice was even seen as a contributor to environmental safety in case of oil spills. Finally, the high costs of energy exploration in iced Arctic waters was used rhetorically as an argument for bringing actors together in cooperative ventures, both nationally and internationally. In other words, Arctic sea ice was also a political resource.

Notes

1. Murmansk on the Kola peninsula was identified as the most suitable option, but as an alternative it was proposed that the pipeline might be extended into northern Norway. This would give the Americans greater control over the envisaged LNG export terminal.
2. In the Beaufort Sea 1975 and 1983 were particularly difficult years.
3. Especially after the 1984 elections.

References

AB Isfjorden-Belsund (1915) 'Rapport öfver AB Isfjorden-Belsunds Spetsbergsexpedition sommaren 1915', vol. 6, AB Isfjorden-Belsund protokoll 1915, Spetsbergenarkivet, Riksarkivet (National archives), Stockholm.

AMAP (2007) *Arctic Oil and Gas 2007* (Oslo: Arctic Monitoring and Assessment Programme, XIII, 40).

Andersson, G. (1917) 'Spetsbergens koltillgångar och Sveriges kolbehov', *Ymer*, 37, 201–48.

Armstrong, T., G. Rogers and G. Rowley (1978) *The Circumpolar North: A Political and Economic Geography of the Arctic and Sub-Arctic* (London: Methuen).

Avango, D. (2004) 'Industriminnesforskning på Svalbard: tolkningar av kulturlandskapet vid Sveagruvan', in Jernkontoret. Bergshistoriska utskottet (ed.), *Arktisk gruvdrift II. Teknik, vetenskap och historia i norr* (Stockholm: Jernkontoret. Bergshistoriska utskottet).

—— (2005a) *Sveagruvan: Svensk gruvhantering mellan industri, diplomati och geovetenskap* (Stockholm: Jernkontoret).
—— (2005b) 'Sveagruvan och Svalbardtraktaten: samarbete och konflikt i kamp om ingenmansland', in A. L. Kemdal, H. Sjunnesson and L. Paulsson (eds), *Daedalus 2005: Tekniska museets årsbok*, 73 (Stockholm: Tekniska museet).
Avango, D. and A. Houltz (2012) '"The Essence of the Adventure": Narratives of Arctic Work and Engineering in the Early 20th Century', in L. Hacquebord (ed.), *LASHIPA: History of Large Scale Resource Exploitation in Polar Areas* (Circumpolar Studies, 8; Groningen: Barkhuis Publishing).
Avango, D. et al. (2008a) 'LASHIPA 3: Archaeological Expedition on Spitsbergen August 7–24, 2006' (Groningen: Arctic Center, University of Groningen).
—— et al. (2008b) 'LASHIPA 4: Archaeological Expedition on Svalbard August 2–25, 2007' (Groningen: Arctic Center, University of Groningen).
—— et al. (2009) 'LASHIPA 5: Archaeological Expedition on Spitsbergen 27 July–17 August' (Groningen: Arctic Center, University of Groningen).
—— et al. (2010) 'Between Markets and Geo-Politics: Natural Resource Exploitation on Spitsbergen from 1600 to the Present Day', *Polar Record*, 47:1, 29–39.
Bach, H. C., and J. Taagholt (1976) *Udviklingstendenser for Grønland: ressourcer og miljø i global sammenhæng* (København: Nyt Nordisk Forlag).
Barr, S. (2001) 'International Research in Svalbard c. 1960–1985: A Cold War Utopia or a Pre-Glasnost Sparring Area?' in Eugene Bouzney (ed.), *International Scientific Cooperation in the Arctic* (Moscow: Institute of Archaeology of RAS).
Brown, R. R. N. (1915) 'Spitsbergen in 1914', *The Geographical Journal*, 46 (July–December), 10–23.
Cadell, H. M. (1920) 'Spitsbergen in 1919', *The Scottish Geographical Magazine*, 36, 1–10.
Coates, P. A. (1993) *The Trans-Alaska Pipeline Controversy: Technology, Conservation, and the Frontier* (Fairbanks: University of Alaska Press).
Curtis, M. I., and D. B. Huxley (1985) 'First Arctic Offshore Field, Endicott, on Decade-Long Way to Production', *Oil and Gas Journal*, 24 June.
Curtis, M. I., and D. B. Huxley (1919) 'Spetsbergenfrågan', *Stockholms Dagblad*, 26 January.
De Geer, G. J. (1912) 'The Coal Region on Central Spitsbergen', *Ymer*, 32, 335–80.
—— (1919) 'Spetsbergenfrågan', *Stockholms Dagblad*, 26 January.
Dole, N. H. (1922a) *America in Spitsbergen – The Romance of an Arctic Coal-mine*, vol. 1(Boston: Marshall Jones Co.).
—— (1922b) *America in Spitsbergen – The Romance of an Arctic Coal-mine*, vol. 2 (Boston: Marshall Jones Co.).
Enström, A. F. (1923a) *Den svenska stenkolsdriften på Spetsbergen* (28 edn, Ingenjörsvetenskapsakademien meddelande; Stockholm: Ingenjörsvetenskapsakademien) 22.
—— (1923b) 'Den svenska stenkolsdriften på Spetsbergen', *Teknisk tidskrift*, 53, 57–62.
Gautier, D. L., et al. (2009) 'Assessment of Undiscovered Oil and Gas in the Arctic', *Science*, 324:5931, 1175–9.
Gustafson, T. (1989) *Crisis amid Plenty: The Politics of Soviet Energy under Brezhnev and Gorbachev* (Princeton: Princeton University Press).

Hacquebord, L., and D. Avango (2009) 'Settlements in an Arctic Resource Frontier Region', *Arctic Anthropology,* 46:1–2, 25–39.
Hartnell, C. (2009) *Arctic Network Builders: The Arctic Coal Company's Operations on Spitsbergen and Its Relationship with the Environment* (Michigan Technological University).
Hemming, A. (1921) 'Möjligheterna för kolskeppningar från Spetsbergen', *Teknisk tidskrift,* 51:7, 73–6.
Högbom, B. (1914) 'Spetsbergens koltillgångar', *Jernkontorets annaler,* Ny serie 69.
Högselius, P. (2013) *Red Gas: Russia and the Origins of European Energy Dependence* (Basingstoke and New York: Palgrave Macmillan).
Johnsson, B. (1916) *Spetsbergens Svenska Kolfält* (Stockholm).
Kaustinen, O. M. (1983) 'A Polar Gas Pipeline for the Canadian Arctic', *Cold Regions Science and Technology,* 7, 217–26.
Lajus, J. A. (2004) 'From Fishing to Mining: The Change of Priorities in the Development of the North and Russian Expeditions to Spitsbergen in the early 20th Century', in Arktisk gruvdrift: teknik, vetenskap och historia i norr. 2, Föredrag presenterade vid ett seminarium på Kungl. Vetenskapsakademien den 6 maj 2000 (Stockholm: Bergshistoriska utskottet, Jernkontoret), 93–106.
Lundström, E. (1920) 'Spetsbergens öde', *Svenska Dagbladet,* 24 February.
Merrit, R. D. (1986) 'Chronicle of Alaska Coal-Mining History' (Fairbanks: Alaska Division of Geological and Geophysical Surveys).
Slavkina, M. V. (2002) *Triumf i tragediya: Razvitie neftegazovogo kompleksa SSSR v 1960–1980-e gody* (Moscow: Nauka).
Wilhelm, C. L. (1919) 'Till Spetsbergen och åter', *Barnens Dagblad,* 21 September.
Von Holstein, L. S. (1920) 'Spetsbergens framtid', *Dagens tidning,* 23 March.

8
Changing Arctic – Changing World

Miyase Christensen, Annika E. Nilsson and Nina Wormbs

The Arctic sea ice is a powerful image of Earth's polar regions. Krupnik et al. (2010) write that frozen saltwater that hardens on the surface of the sea is a crucial component of the global system and life in the polar regions. The description of its nature by scientists and indigenous observers brings this vision of the ice even closer to the heart: 'it can be stiff and silent but also blasting and crushing with terrible noise, and can advance and retreat as a living being' (ibid., p. 2). Today, Arctic sea ice is not only a significant element in scientific discourse and local knowledge, but a key constituent of global climatic and social change.

The 2007 Arctic sea-ice minimum served as the starting point for this volume on media and the politics of Arctic climate change. The chapters have addressed various discourses that materialized around this theme along a past-to-present continuum. As we write at the end of 2012, a new record minimum was set. On 16 September 2012, the Arctic sea-ice extent averaged only 3.41 million square kilometers, well below the 2007 figure, which was the previous record low, and 3.61 million square kilometers below the 1979-to-2000 average extent (see NSIDC. org).[1] Some of the improved models of Arctic sea ice suggest that the Arctic Ocean could be almost ice-free in the summers in only 30 years (Wang and Overland, 2012).

Following a trend of framing the Arctic as bellwether of climate change, as presented in Chapter 2 of this volume, the quality press continued to publish stories about the significance of sea-ice loss and resulting climate change as a *meta-event*, emphasizing scientific certainty and using alarmist metaphors. In a *New York Times* story (September 2012), Walt Meier from the National Snow and Ice Data Center was quoted as suggesting 'The Arctic is the earth's air-conditioner. We're losing that.

It's not just that polar bears might go extinct, or that native communities might have to adapt, which we're already seeing – there are larger climate effects.'[2] The story also cited the NASA climate scientist James E. Hansen saying: 'The scientific community realizes that we have a planetary emergency.' Similarly, *The Guardian* published a story entitled 'Arctic ice shrinks 18% against record, sounding climate change alarm bells.'[3] The story made reference to author and environmental campaigner Bill McKibben, who commented: 'Our response [so far] has not been alarm, or panic, or a sense of emergency. It has been: "Let's go up there and drill for oil". There is no more perfect indictment of our failure to get to grips with the greatest problem we've ever faced.'

While we should remain cautious of making overarching assumptions about media's coverage of Arctic climate change (particularly across tabloid and quality press; web-based media and television; and national cultures), our own studies and the work of others on discursive patterns indicate that media's treatment of climate change increasingly highlights broader questions of global change (see also Boykoff and Roberts, 2007). Moreover, the range of topics that become connected to Arctic sea ice in the media show how climate change has become an issue relevant not only for climate scientists but also for society at large, making it a key issue for humanities and social sciences.

The express purpose of this volume has been to address key aspects of Arctic climate change from an interdisciplinary perspective and in the context of our highly mediatized late-modern era. The findings presented in the various chapters point to a number of key issues that surpass the single phenomenon of measurable, observable 'change' in the Arctic sea ice. Specifically, the sea-ice minimum and the scientific, political and public discourses that materialize around it have overarching implications regarding the dialectical relationship between human-induced climate change, globalization processes and social transformation.

This chapter concerns itself with a number of broader questions raised by the Arctic sea-ice minima. The media dimension serves as an entry point to discuss social trends and the role of humanities and social sciences in the overarching context of global change.

Climate change between science and the media

Anderegg et al. (2010) report that there is 97 percent agreement among actively publishing climate scientists that human-caused global warming exists. Yet there can be, and oftentimes is, great disjuncture between what reality *is* and the media story it *becomes*. The variety of factors that underlie such disjuncture range from the profit-oriented political economy of media institutions to journalistic norms of objectivity and balance, and the perceived need to dramatize news stories so as to compete with other high-drama events (for an extended discussion see Boykoff and Roberts, 2007; Christensen, 2011).

A recurring pattern in journalistic coverage is that norms of objectivity and the principle of 'balance' often reign supreme and compromise other aspects of responsible news reporting. When news reporters dedicate equal media space and time to climate skeptics, it inevitably gives the impression that scientific opinion on the matter is equally divided, despite the wide-scale consensus within the scientific community as to the existence and causes of climate change. One study surveying 636 articles from top-ranked US newspapers between 1988 and 2002 found that a clear majority of the articles devoted as much space to the marginal group of climate-change skeptics as they did to those representing scientific consensus on the issue (Boykoff and Boykoff, 2004; see also Oreskes and Conway, 2010). Such reporting has consequences. One such example is how the online release of e-mails from climate scientists at the University of East Anglia turned into a media event: 'Climategate'. Polls suggest it contributed to many in the United States losing faith in climate scientists (Leiserowitz et al., 2010; Marquart-Pyatt et al., 2011) and created small shifts towards more skeptical positions in the United Kingdom.[4]

Quantitatively, the media coverage of climate change follows a pattern of story peaks around 'events' and 'moments' of media significance, followed by dips in the attention cycle. September, being the month of annual minimum measured in Arctic sea-ice extent, constitutes one such (recurring) moment. Particularly since 2007, there has been an increase in the news stories and reports about Arctic climate change around this time, with due space allocated in news outlets anticipating the announcement of newsworthy facts and figures. This and other peaks fall within an overall trend of increase in the media attention paid to climate change over the past decade whereas 2010 brought a significant decline in the coverage of climate change in most parts of the world – in some cases to pre-2007 levels (Boykoff and Mansfield, 2010). The Daily Climate (dailyclimate.org), a website that tracks media coverage of climate change, concluded that '2010 was the year climate coverage fell off the map'. Their database showed that the volume of coverage in English-language publications dropped by 30 percent to 2005 levels. There is also research indicating that for the whole of 2010, TV coverage in the United States was minimal.[5]

The drop can be largely attributed to lack of attention to climate-change-related issues in political circles (particularly on Capitol Hill), but further research would be needed to establish the actual causes.[6]

Regarding the quality of reporting, research shows that coverage of physical-science aspects of climate change has improved in the last decade: 'the press has been quite reformist in its portrayal of the needed action on climate change, when the scientific projections suggest the issue may call for truly revolutionary changes', (Boykoff and Roberts, 2007, p. 34). While in the decades preceding the 2000s the media focused on technical aspects and mitigation, and a disproportionally represented sense of scientific controversy, the past decade of news reports have featured more and more stories that address political, social, cultural and ethical aspects of climate change, including the question of adaptation. It certainly will take further research to conclude whether there is such a trend across the board. Yet it is our contention that the quality press has been paying more attention to the social and cultural dimensions of climate change throughout the 2000s, particularly after 2007. The scientific foundation is there. Not only did the 2007 IPCC report confirm the human factor behind climate change, but there was increasing attention paid to impacts and associated costs (IPCC 2007a, 2007b). On the whole, as we discussed in Chapter 2, these combined factors, such as the passing of the era of laissez-faire globalism (following environmental and financial disasters) and the rising significance of sociocultural aspects, provided linkages between different topics covered within the same media narrative – *topical multiplicity* – as well as the linking of global and local issues – *scalar transcendence*. There was also a more pronounced emphasis on elements such as scientific certainty.

The moral of the story

News media are indispensable to the construction of public understanding regarding the relationship between human life and the natural and built environment. In spite of recent debates on the linkages between volume of coverage and individual/public understanding towards particular issues, it is reasonable to suggest that the media matter a great deal for how society understands climate change today, including its multiple social and cultural dimensions. However, important ethical and moral dimensions, especially those that include responsibility for the future – rather than value-neutralism – have appeared to be missing, particularly in the US media. There are many reasons for this. First, global climate change as a scientific concept has for long been too abstract and complex for the purposes of news-making. Journalists prefer 'morality plays' with clear and concrete winners and losers and rights and wrongs. Secondly, and as we discussed earlier, journalistic norms of balance, truth and objectivity have been impediments to capturing the scientific discussions accurately. Finally, a concept such as climate change makes for a better media story if it involves controversy (i.e. drama). In the absence of concrete story elements that can portray the moral dilemmas of climate change, the rights and wrongs have become projected onto the practice of science.

Environmental disasters in the late 1980s provided dramatic moments for the media to generate alarmist discourses raising questions about the linkages between such events and global warming/climate change. Morality was also apparent in the coverage of climate politics. For example, reporting on international protocols such as Kyoto involved casting of key political figures and heads of state such as Bush along the lines of 'good guys' versus 'bad guys'.

The visible impacts of climate change in connection with the 2007 sea-ice minimum provided a new moral dimension, one that went beyond local disasters and challenged previous perceptions of our planet as being ultimately resilient and impervious to human damage. Parallels could be drawn with the discovery of the ozone hole over Antarctica in the mid-1980s, which shattered the old conviction that human pollution was irrelevant on the global scale. It may be going too far to claim that Arctic sea-ice minima instigated a moral turn, on the whole, in the international news coverage of climate change. Yet it clearly provided that concreteness and high-drama moments that the media crave, including the tragic sublime embodied in the form of giant chunks of ice crumbling into the sea and starving polar bears left with nothing but a piece of ice to float on. Combined with clear messages from the IPCC and the scientific community that Arctic change was a testimony to human impact on the global climate, it became easier for journalists to cover both the local and global dimensions of climate change – including the high social, cultural and economic stakes – which had been largely missing from the stories published throughout the 1980s and 1990s. Climate change could simply no longer be regarded as a mere question of data sets and scientific turf wars. The multi-topical dimensions of global climate change, ranging from energy, food and health to local communities and farmland and to vulnerable populations and need for adaptation, had become all too apparent (IPCC, 2007a). Similarly, the impacts on Arctic landscapes and people documented in the Arctic Climate Impact Assessment (ACIA, 2004), and the rapidly declining Arctic sea ice charted in 2007 also played a significant role. Covering this multiplicity and its corresponding geographic consequences necessitated dedicating human resources such as reporters and commentators capable of dealing with such scientific, political and economic complexity. As evinced by the media coverage, quality press (as well as small media and local groups) responded to this challenge. More voices could be heard, including voices of those directly affected by climate change, not least in the Arctic, resulting in increasing attention to the ethical dimension of climate change. Some of those voices gained power from the increasing scientific attention to Arctic climate change (Nilsson, 2007). Whether the moral trend will be sustained and effective (despite routine drops in media attention) and how it links with civic action and political intervention remains to be seen.

Climate change and social theory in an era of complex mediation

The significance of mediated information for public understanding, public policies, civil action, cultural practice and political discourse makes the coverage of climate change and Arctic sea ice relevant across the disciplines of humanities and social sciences. Environmental change informs our understanding of social change on the whole, the ways in which we think about the direction of change and our role in determining that directionality. Globalization, as a paradigm, has moved from more laissez-faire technocratic and culturally deterministic understandings towards a moral discourse of environmental damage, human cost, controlled change and global governance (for example Biermann et al., 2012). This trend has been particularly visible in the wake of multiple financial crises – which are equally global in scale and scope as climate change. At the same time, geopolitical concerns over territory and resources become apparent political priorities, not least in the Arctic (Huebert et al., 2012). In order to understand and address the issues at stake in all their complexity, it becomes more and more necessary to incorporate perspectives across scientific disciplines within the realms of both natural sciences and social sciences and humanities.

The chapters that constitute this collection and the vision we promote are one attempt at such a cross-cutting approach. The research project that brought us together is characterized by what the funding agency has regarded 'interdisciplinary research'. Yet interdisciplinary research itself can be defined in different, contested, ways and depends partly on how one defines a discipline and constructs patterns of connectivity.

Increased specialization within knowledge-production regimes might mean that areas that used to be part of the same are today organizationally, methodologically or intellectually different, constituting autonomous disciplines. New areas of research continuously take shape bringing new disciplinarities while old ones wither, both within natural sciences and social sciences and humanities.

Historian and geographer David Lowenthal has argued that the increasing specialization and distance between disciplines are detrimental to the unity of knowledge. We can no longer turn to the experts (Lowenthal, 2012a). Not only are their stories increasingly difficult to understand because of specialized language, but different experts also present singular truths about the world which are hard to merge into a coherent picture of the complex socio-economic and cultural processes of the twenty-first century.

The oft-articulated solution involves trans-, multi-, inter- or perhaps even post-disciplinary research and training of new types of scholars who are able to write stories that are intelligible across the board, spanning more than just a specific field. It goes without saying that such synthesis never comes easy and it requires special measures to facilitate post-disciplinary fusion. Funding structures and academic career systems based on disciplinary specialization poses obstacles. To play outside of the established institutional contexts is risk-taking because the academic career system is still mainly based on disciplinary quality criteria and because it takes time to establish good interdisciplinary dialogue in research projects, which can affect the quantity of published results (Østreng, 2009).

We need to find ways of overcoming these obstacles. Climate change and adaptation, poverty and social injustice, famine and health care are issues too big to be solved within disciplinary boundaries. We need as much knowledge about social and cultural processes as we need about physical aspects of the environment as we continue to form stories, narratives and understandings of events (within academia as well as in public discourse) because these could ultimately affect and guide collective and individual action. In philosophy, history or literature, there have been strands of scholarship focusing on the *environment*, even before the term was used in its contemporary sense. Now environmental philosophy, environmental history and ecocriticism are gathered under the umbrella of environmental humanities, reflecting a need to bridge knowledge production within the humanities. Similarly, environmental social sciences have been pulled together within larger research programs, such as the different activities commenced under the International Human

Dimensions Programme. The latter is currently being morphed together with the natural-science-focused International Geosphere-Biosphere Programme into a new venture called Future Earth.

Particularly within the last decade, environmental communication research has also matured into an established field (Hansen, 2011). International research associations such as IAMCR (International Association of Media and Communication Research), ECREA (European Communication Research and Education Association) and ICA (International Communication Association) created new divisions and working groups around the study and research of climate and the environment. Development in the Arctic as portrayed in this book and in the increasing social science and humanities literature on Arctic change shows that polar regions, as a whole, serve as a useful focus area for multidimensional analysis and an entry point towards discussing how we are to understand more fundamental changes in the world in which we live.

From Gaia to the Anthropocene

As indicated before, the Arctic has often been referred to as the bellwether of climate change (ACIA, 2004; Nilsson, 2007). It has been used to signify that something remarkable is happening to the global environment in ways that become particularly visible in the Arctic. Similar metaphors have been used in relation to other environmental problems, pollution in particular, where the Arctic environment magnifies impact on people and nature (AMAP, 2002). More recently, the bellwether metaphor has been extended to the social domain and invoked for describing security politics in the Arctic (Huebert et al., 2012). Regardless of how the biophysical and social processes are linked, it is fair to say that Arctic change, as it is playing out at the moment, is an illustration of Earth having moved into a new geological era that has been called the Anthropocene (Crutzen, 2002).

As the term implies, the most significant feature of this era is that the environment is shaped by human activities not only at the local scale, but also at a global level. This era replaces the Holocene, which has featured an unusually stable climate and has been the period during which human society grew from small and scattered populations to the large, organized societies of the Industrial Era. Today, our globalized society is characterized by increasingly dense connections in a range of intersecting systems. Not only are markets and the media globalized, but the same holds true for the flow of natural resources – as raw material or in the form of consumer products. Global distribution of byproducts or emissions from human activities is another facet, including not only emissions of greenhouse gases but also ozone-depleting chemicals and some toxic pollutants such as persistent organic pollutants and mercury. Human activities have also had large-scale impact on the biogeochemical cycles of water, nitrogen and phosphorus, which are all essential resources for food production.

In the 1970s, the physicist James Lovelock wrote about the planet Earth in relation to what he termed the Gaia hypothesis. The main idea was that Earth is a self-regulating system that will seek to maintain equilibrium, initially including an implicit assumption that human activities were insignificant in relation to Earth's homeostatic mechanisms. It was criticized already during its time, but suffered a more severe blow due to the observations of the Earth's atmosphere, which yielded the discovery of the ozone hole over Antarctica in 1985. Chemicals produced by humankind did indeed change the atmosphere in ways that affect life on Earth in fundamental ways. The evidence of anthropogenic climate change also shows that the magnitude of power human society has over nature is large enough to be conceived as a *geological force*.

More recently the discourse on global change has incorporated not only the notion of major gradual change, but also attention to thresholds or *tipping points* (Lenton et al., 2008). These are abrupt changes in biophysical or social systems that can be the result of many interacting drivers of change. The term tipping point is often, but not always, used to describe a change that is difficult to reverse. It is a regime shift into a new stable state. Tipping points have been studied for some time in local ecosystems, such as lakes turning from clear to turbid. In the Arctic, there is increasing interest in understanding the risk for thresholds in both ecosystems and in linked social-ecological systems (Wassmann and Lenton, 2012). One could argue that the loss of Arctic sea ice in the summer, in and of itself and in terms of its consequences, is an example of the Arctic having passed an irreversible threshold. However, given the complexity of the systems involved, we may not fully grasp for many years the significance of what is happening now.

The notion of tipping points also appears as an underlying assumption in the increasingly popular concept of *planetary boundaries*. This concept was coined by Rockström et al. (2009) as an attempt to visualize the vulnerable situation of the global environment as a whole – the Anthropocene version of the Gaia hypothesis where the homeostasis is lost because of large-scale human impacts on the environment. The notion that self-regulating mechanisms are no longer functioning fits well with increasing interest in discussing global governance and how it can be strengthened (Biermann et al., 2012).

There is no doubt that global governance has failed to keep pace with the increasing impacts of human activities on the environment. It is clear that environmental problems will make it difficult for large numbers of people to fulfill their basic needs for food, drinking water and sanitation. On the basis of the failure of climate governance to reach agreement on reducing greenhouse gases, one should also question the lack of political willingness to act. In the Arctic there is a race for fossil-fuel resources where the actors do not appear to question the morality of contributing to further emissions of greenhouse gases. In addition, the Arctic environment has gone from a field of low politics amenable to cooperation in the 1990s, to one of high politics, with increasing attention to issues of national security after 2007 (Nilsson, 2012). The media's role in this shift is significant.

In sum, the importance of the Arctic goes well beyond climate change and its environmental impacts. The mediatized discourses are indicative of the scalar and topical complexity of the challenges that mark today's global society at a fundamental level. Change and continuity in the social sphere has been the subject of inquiry in social theory in numerous contexts. A new challenge has been the incorporation of environmental change and its material and symbolic dimensions to the discourse of globalization and social change.

From progress to open-ended futures

Public culture and scholarly production influence one another in terms of the ways in which certain values and phenomena are regarded at a given time and place. This also pertains to perceptions and experiences of the environment and environmental crisis. Going back to the heyday of Western sociology, Beck (2010, pp. 255–6) makes note of a pertinent quotation from Max Weber: 'bis der letzte Zentner fossilen Brennstoffs verglüht ist' ('until the last ton of fossil fuel has burnt to ashes'). Pointing to the ecological subtext, Beck added: 'This is more than merely a metaphor. In Weber's view, industrial capitalism generates an insatiable appetite for natural resources which undermines its own material prerequisites.' While the environment has always been a recurring theme in human thought and cultural production from philosophy to art, the environmental movement of the 1960s and 1970s opened up new avenues of thinking, a search for new epistemologies in the humanities and social sciences as well as increasing popular and academic interest in environmental change. Such epistemologies went well beyond natural-scientific conceptions of the environment.

From the 1990s onward, the discussions have included two new discursive frames that have since occupied the research agendas of the social sciences and humanities. The first was the attention paid to the spatial extension of networked capitalism often shorthanded as 'globalization', which gained momentum in the past two decades. The second was intensifying media saturation facilitated by a growing sector of digital technologies. These shifts have generated a large body of scholarship, part of which was theoretically sound, sharp and inspiring (Silverstone, 1994, 2006; Sassen, 2008; Balibar and Wallerstein, 1991; Morley and Robins, 1995). Some were of more speculative nature, yet yielded further questions (Fukuyama, 1992; Barber, 1996; Giddens, 1990; Appadurai, 1996). In social theory, while a large amount of intellectual space was dedicated to cultural consequences and virtual flows, explorations of the material and environmental consequences of global capitalism and consumerism have remained limited. In the latter half of the 1990s and particularly in the 2000s, 'climate change' occupied significant space in public debates and popular imagination (such as film). This has been met with a corresponding increase of interest in incorporating the environment as an area of focus in both social science curricula and research.

In social theory, there have been some notable examples of this trend of environmental focus which are both widely influential *and* contested. Names include scholars such as Ulrich Beck, John Urry, Anthony Giddens and John Gray (to name but few), all of whom have attempted to build linkages between ecological issues and social theory (for example, what does climate change reveal about modernity and capitalism?), albeit from varied political and epistemological standpoints. In his 1992 book *Risk Society*, Beck framed environmental problems and economic inequality as the 'bads' or the negative outcomes of modernity while still seeking the solution in modernity itself, and arguing against a nature–culture divide. For Beck, risk society is, in essence, what defines the state of affairs in late modernity. In his discussion, Beck placed particular emphasis on environmental destruction and the ensuing catastrophic consequences for human society (see also Barry, 2007). At the individual level, the mediated understanding of the volatile nature of the global economy, financial markets, political regimes, cultural borders, and the fact that we share a global destiny with an ultimately unstable climate marks the lifeworld in unique ways. Such an understanding of the illusiveness of historically linear progress and human prosperity (tenets of first modernity) is also indicative of a loss of faith in 'progress' in general, and in sociopolitical institutions and their grand rhetoric of 'restoration of social order and moral and material wealth' in particular (Christensen and Jansson, forthcoming). As Beck and Willms note (2004, p. 34): 'Not only is the future indeterminate, but its indeterminacy is part of the meaning of present.'

The theoretical and ideological underpinnings of Beck's position on politics and environment map onto the trope that constitutes Gidden's 'new left' vision as they both seek 'post left–right' political solutions to current social problems and increasing risks. Giddens promotes an alarmist discourse and *climate change New Deal* across nations towards economic and environmental benefits. He declares, in a recent newspaper piece entitled 'Recession, Climate Change and the Return to Planning', that the era of deregulation is over and 'the state is back', pointing a finger at globalization: 'And then, well, there's the granddaddy of the whole thing, globalization, which has proceeded apace without adequate international controls.'[7] At the other end of the spectrum, Gray's eco-conservative posthumanism and disbelief in the moral autonomy of humans stand in stark contrast to left-leaning schools of thinking in social theory which see human agency as the primary condition for human progress and social development. Gray's vision, rooted in Lovelock's Gaia hypothesis and Edmund Burke's conservatism, sees progress as an impossibility and political intervention as doomed utopianism. It goes without saying that social theory has engaged issues of climate change for a long time and the perspectives and thinkers noted here offer just a glimpse of the accounts which have been particularly influential in driving intellectual agendas of the field and prompting controversy.

The frozen saltwater that solidifies on the sea surface depicted in the opening paragraph of this chapter is both immediately material in its physical qualities and a powerful image. It is part of a seascape that has shaped human activities in the Arctic for centuries, limiting some and enabling others. Together with other features of the Arctic, the ice continues to play an important role in shaping our material reality and social imaginary. Social theory, just as other areas in humanities and natural sciences, is both abled and bounded by its own epistemological connects and disconnects, paradigmatic trends and disciplinary reflexes. It is no small feat to break out of this pattern, even if there is a push towards interdisciplinary and innovative agendas through new funding schemes and institutional re/organization. The transcendent scale and multivalent scope of the greatest challenge we are faced with today, climate change, necessitates a truly outward and forward-looking

Changing Arctic – Changing World 169

post-disciplinarity eager to connect and not afraid to reroute and re-try when dead ends emerge along the way. Such an approach is essential not only to grasp the crucial issues at stake and produce cutting-edge research, but to informing global–local policy agendas and international governance.

Notes

1. National Snow and Ice Data Center, http://nsidc.org/arcticseaicenews/2012/09/arctic-sea-ice-extent-settles-at-record-seasonal-minimum/, date accessed 16 February 2013.
2. *New York Times*, http://www.nytimes.com/2012/09/20/science/earth/arctic-sea-ice-stops-melting-but-new-record-low-is-set.html?_r=0, date accessed 16 February 2013.
3. *Guardian*, http://www.guardian.co.uk/environment/2012/sep/19/arctic-ice-shrinks, date accessed 16 February 2013.
4. *Guardian*, http://www.guardian.co.uk/environment/2010/jul/07/climate-emails-question-answer, date accessed 16 February 2013.
5. *Daily Climate*, http://wwwp.dailyclimate.org/tdc-newsroom/2011/01/climate-coverage, date accessed 16 February 2013.
6. Methods used in counting and the reliability of online archives are also factors that need to be considered.
7. *Huffington Post*, http://www.huffingtonpost.com/anthony-giddens/recession-climate-change_b_174424.html, date accessed 17 February 2013.

References

ACIA (2004) *Impacts of a Warming Arctic: Arctic Climate Impact Assessment* (Cambridge: Cambridge University Press).
AMAP (2002) *Arctic Pollution Issues 2002* (Oslo: Arctic Monitoring and Assessment Programme).
Anderegg W. R. L., J. W. Prall, J. Harold and S. H. Schneider (2010) 'Expert Credibility in Climate Change: Proceedings of the National Academy of Sciences, USA', 107, 12107–09.
Appadurai, A. (1996) *Modernity at Large: Cultural Dimensions of Globalization* (Minneapolis and London: University of Minnesota Press).
Balibar, E., and I. Wallerstein (1991) *Race, Nation, Class* (London: Verso).
Barber, B. (1996) *Jihad vs McWorld: How Globalism and Tribalism are Reshaping the World* (New York: Ballantine).
Barry, J. (2007) *Environment and Social Theory* (London: Routledge).
Beck, U. (1992) *Risk Society: Towards a New Modernity* (London: Sage).
Beck, U. and J. Willms (2004) *Conversations with Ulrich Beck* (Cambridge: Polity Press).
Beck, U. (2010) 'Climate for Change, or How to Create a Green Modernity?' *Theory, Culture & Society*, (March/May), 27:2–3, 254–266.
Biermann, F. et al. (2012) 'Navigating the Anthropocene: Improving Earth System Governance', *Science*, 335:6074 (16 March), 1306–7.

Boykoff, M. T. and J. M. Boykoff (2004) 'Balance as Bias: Global Warming and the US Prestige Press', *Global Environmental Change*, 14, 125–36.

Boykoff, M. T. and J. T. Roberts (2007) *Media Coverage of Climate Change: Current Trends, Strengths, Weaknesses* (Oxford: UNDP).

Boykoff, M. T. and M. Mansfield (2010) *Media Coverage of Climate Change/Global Warming*. Center for Science and Technology Policy Research, University of Colorado and University of Exeter, Oxford University. Online: http://sciencepolicy.colorado.edu/media_coverage/.

Christensen, M. (2011) 'Discursively Shaping the Environment: Swedish National and Regional Media Coverage of The United Nations Climate Change Conference in Copenhagen ("COP15")'. Paper presented to the Global Communication and Social Change Division of the International Communication Association (ICA) Conference, May 2011, Boston.

Crutzen, P. J. (2002) 'Geology of Mankind', *Nature*, 415:23.

Fukuyama, F. (1992) *The End of History and the Last Man* (New York: Free Press).

Giddens, A. (1990) *The Consequences of Modernity* (Cambridge: Polity).

Hansen, A. (2011) 'Communication, Media and Environment: Towards Connecting Research on the Production, Content and Social Implications of Environmental Communication', *International Communication Gazette*, 43:1–2, 7–26.

Huebert, R., H. Exner-Pirot, A. Lajeunesse and J. Culledge (2012) *Climate Change & International Security: The Arctic as a Bellwether*. Online: http://www.c2es.org/publications/climate-change-international-arctic-security.

IPCC (2007a) Climate Change 2007: Climate Change Impacts, Adaptation and Vulnerability. Summary for Policymakers. Working Group II Contribution to the Intergovernmental Panel on Climate Change Fourth Assessment Report (Geneva: Intergovernmental Panel on Climate Change).

—— (2007b) Climate Change 2007: The Physical Science Basis. Summary for Policymakers. Contribution of Working Group I to the Fourth Assessment Report of the Intergovernmental Panel on Climate Change (Geneva: Intergovernmental Panel on Climate Change). Online: www.ipcc.ch/pdf/assessment-report/ar4/wg1/ar4-wg1-spm.pdf.

Krupnik, I., C. Aporta, S. Gearheard, G.J. Laidler, and L. Kielsen Holm (eds.) (2010) *SIKU: Knowing Our Ice Documenting Inuit Sea Ice Knowledge and Use*. Dordrecht: Springer.

Leiserowitz, A. et al. (2010) 'Climategate, Public Opinion, and the Loss of Trust' (July 2). Online at SSRN: http://ssrn.com/abstract=1633932 or http://dx.doi.org/10.2139/ssrn.1633932.

Lenton, T. M. et al. (2008) 'Tipping Elements in the Earth's Climate System', *Proceedings of the National Academy of Sciences*, 105:6 (February 12), 1786–93. doi:10.1073/pnas.0705414105.

David Lowenthal (2012) 'The Quest for the Unity of Knowledge', The Stockholm Archipelago Lectures, KTH Environmental Humanities Laboratory, 19 September.

Marquart-Pyatt, S. T., et al. (2011) 'Understanding Public Opinion on Climate Change: A Call for Research', *Environment: Science and Policy for Sustainable Development*, 53:4, 38–42.

Morley, D., and K. Robins (1995) *Spaces of Identity* (London: Routledge)

Nilsson, A. E. (2007) 'A Changing Arctic Climate: Science and Policy in the Arctic Climate Impact Assessment' (Linköping, Sweden: Department of Water and Environmental Studies, Linköping University).

—— (2012) *The Arctic Environment from Low to High Politics*. Arctic Yearbook 2012. Northern Research Forum and University of the Arctic. Online: http://www.arcticyearbook.com/.

Østreng, W. (2009) (ed.) *Transference. Interdisciplinary Communications 2008/2009* (Oslo: CAS).

Rockström, J., et al. (2009) 'Planetary Boundaries: Exploring the Safe Operating Space for Humanity', *Ecology and Society*, 14:2, 32. Online: http://www.ecologyandsociety.org/vol14/iss2/art32/.

Sassen, S. (2008) *Territory, Authority and Rights: From Medieval to Global Assemblages* (Princeton: Princeton University Press).

Silverstone, R. (1994) *Television and Everyday Life* (London: Routledge)

—— (2006) *Media and Morality: On the Rise of the Mediapolis* (London: Polity).

Wang, M., and J. E. Overland (2012) 'A Sea Ice Free Summer Arctic Within 30 Years: An Update from CMIP5 Models', *Geophysical Research Letters*, 39:18 (September 25). doi:10.1029/2012GL052868. Online: http://www.agu.org/pubs/crossref/pip/2012GL052868.shtml.

Wassmann, P., and T. M. Lenton (2012) 'Arctic Tipping Points in an Earth System Perspective', *AMBIO*, 41:1 (February 1), 1–9. doi:10.1007/s13280-011-0230-9.

Index

ACIA, *see* Arctic Climate Impact Assessment (ACIA)
Advanced Synthetic Aperture Radar (ASAR) instrument, 57
Africa, 2, 66–7, 117
Ahlmann, Hans Wilhelmsson, 76–7, 81, 84, 87n12
Aiken, Wesley, 117
air-conditioner, Arctic as, 157–8
"alarmist" metaphors, 29, 157–8, 161, 168
Alaska, 20, 27, 39, 42–3, 46–7, 115, 117–21, 124, 130–3, 138–9, 143–5
Amundsen, Roald, 78
"An Ice-free Arctic is Completely Possible" (2006) (*Dagens Nyheter*), 40
An Inconvenient Truth (film), 7–8, 30, 100
Annual Meeting of the Royal Academy of Sciences (Stockholm), 81
Antarctica, 65–7, 82, 89n30, 104, 161, 165
Anthropocene, 21, 70–1, 85, 93, 164–6
anthropogenic influence, 70–1, 83–5, 93, 96–9, 102–4, 106, 108–10, 158–66
AOSB, *see* Arctic Ocean Sciences Board (AOSB)
Apollo 8, 53
Apollo 17, 53
Aporta, C., 157
Arab Spring, 4, 6
Arctic, map of, 118
Arctic, as possible sphere center, 66–7
Arctic and Antarctic Research Institute, 81
Arctic Basin, 118
Arctic Climate Impact Assessment (ACIA), 17, 99–106, 108–9, 116, 161–2, 164
Arctic Climate System Study, 16
Arctic Coal Company, 142
Arctic Council, 14–15, 17, 101, 106, 116
Arctic Dipole anomaly, 97, 110n1
Arctic energy exploration (1898– 1985), 128–54
and drilling, 131–3, 143–54
and geopolitics, 144–5, 153
in historical perspective, 129–34, 153
and "iceberg management," 148–50
and jetties, 136
and mining, 129–31, 134–43
and nationalism, 140–1
and oil, gas and ice, 143–52
and pipeline system, 143–6, 151, 154
and "reverse oil price shock," 152–3
and rhetoric, 137–40, 152, 154
and sea ice in the era of coal, 134–43
and snow, 137–9
and sparse population, 151
and storage, 134–7
and wood, 137
Arctic Environmental Protection Strategy, 14–15
"The Arctic: here today gone in 40 years" (2006) (*Guardian*), 41
Arctic Ice (*L'dyArktiki*) (1945) (Schmidt), 76
"Arctic Melt Opens Northwest Passage" (2007) (*National Geographic News*), 61–2
"Arctic Melt Unnerves the Experts," (2007) (*New York Times*), 61
Arctic Ocean Sciences Board (AOSB), 105
Arctic Oscillation, 97, 110n1
Arctic Sea Ice Conference, 80
Arctic sea ice melting, *see* sea-ice minimum (Arctic) (2007)

173

"Arctic Sea Ice Melting Faster, a
   Study Finds" (2007) (*New York
   Times*), 40
"Arctic sea ice 'melts to all-time
   low'(2005 and 2007) (NASA)
   (image), 59–64
"Armageddon," 41, 46
Armstrong, Terence, 79
ASAR, *see* Advanced Synthetic
   Aperture Radar (ASAR)
   instrument
AS17-148-22727 (Blue Marble photo)
   (1972), 53, 65–7
attribution, and climate change,
   95–110
Avsyuk, Grigory, 80

Baffin Island, 117–19
Barents Sea, 74–6, 118, 133,
   146–7, 153
Barrow, Alaska, 115, 117–18,
   120–2, 124
Beaufort Sea, 118–19, 131–3, 147–8,
   150, 153, 154n2
Beck, Ulrich, 166–8
Belgium, 56, 146
bellwether of climate change, 21n5,
   101, 103, 107, 157, 164
Bent, Silas, 72
Bergen school of meteorology, 76–7
Bering Strait, 119, 123, 144, 147–8
Berlin Wall, 6
Bjerknes, Vilhelm, 77
Blue Marble photo (original) (1972),
   53, 65–7
Blue Marble photos, 53, 63–7
   and the blurring of data sources,
      64–6
   and Eastern Hemisphere, 64
   original (1972), 53, 65–7
   new (2012), 64–6
   and Western Hemisphere, 64
BOREAS program, 85
"breaking news," 5, 70
   see also drama, and media
Brooks, C. E. P., 72
Burke, Edmund, 168
Bush, George W., 41–2, 45–6, 161
Byrnison, Iorek, x

Cadell, Henry, 138
Canada, 46, 79, 83, 115, 117–18, 131,
   145–6, 148, 150, 152
Canadian Arctic Archipelago, 117–18
capitalism, global, 21, 43–4, 166–7
censorship, 42–3, 46
Central Intelligence Agency
   (CIA), 80
CERN, *see* European Organization for
   Nuclear Research CERN
"Climategate," 159
China, 21, 40, 42, 46, 153
Chukchi Sea, 118–21, 124
Chukotka, Russia, 27, 117–18, 121
CIA, *see* Central Intelligence Agency
   (CIA)
climate change and sea-ice, *see* sea-ice
   minimum (Arctic) (2007)
"climate change New Deal," 168
climate change "skeptics," 6,
   41–2, 159
"climate change" versus "global
   warming" (U.S.), 20, 47
climate variability, 20, 47, 93–110
Climatic Research Unit, 8
coal, 20, 128–43, 153–4
Cold War, 15–17, 58, 63, 71, 77,
   79–86
commercialization, 1–3, 14, 55–6,
   73–4, 123, 131–2, 148, 152
Copenhagen climate summit (2009)
   (COP15), 5, 8, 30
COP15, *see* Copenhagen climate
   summit (2009) (COP15)
Crutzen, Paul, 85

Dagens Nyheter (*DN*) (Sweden), 19, 27,
   32–8, 40, 48
   pieces published (2003–2006), 35
   pieces published (2007–2010), 35
   pieces published (2003–2010), 36
   sectional breakdown of (2003–
      2010), 37
Daily Climate website, 159
DAMOCLES, 16
Danish Meteorological Institute, 75
Dayan, D., 4
De Geer, Gerard, 140–1
Denmark, 27, 79, 148

detection, and climate change, 95–9, 104, 106
"The disappearing world of the last of the Arctic hunters," (2010) (*Guardian*), 44
Dome Petroleum, 151
drama, and media, 1–5, 30, 34–5, 41, 70, 79, 107, 114, 118, 125, 131, 159, 161
drilling, 43, 123, 131–3, 143–54
drought, 5, 30, 38–9, 72

"Early Arctic Warming" period (1919–1940s) (EAW), 72
Earthrise picture (Apollo 8) (1968), 53
Eastern African coast, 2
EAW, *see* "Early Arctic Warming" period (1919–1940s) (EAW)
ecosystems, 27–8, 107, 117, 165
ECREA, *see* European Communication Research and Education Association (ECREA)
Edgeøya, 146
Endicott, 131
energy, *see* Arctic energy exploration (1898–1985); natural resources
Envisat satellite, 57
ESA, *see* European Space Agency (ESA)
"Eskimos Seek to Recast Global Warming as a Rights Issue" (2004) (*New York Times*), 44
Esrange Space Center, 56
European Arctic warming trend (1920s), 73–4
European Communication Research and Education Association (ECREA), 164
European heat wave (2003), 5
European Organization for Nuclear Research CERN, 56
European Science Foundation BOREAS program, 85
European Space Agency (ESA), 56–7, 61–3
Europe, 5, 16, 54–7, 61, 72–4, 77, 85, 87n12, 105, 129, 132–4, 138, 140, 142, 146, 164
European Union (EU), 16, 105
eventization, 2, 6, 159

Exxon, 131

feedback, 16, 85, 97, 101, 103, 106–7
feedback loops, 16–18, 41
"fender platform," 148–50
first-year ice, 18, 70
Floating Maritime Research Institute, 74
fossil fuels, 20, 73, 83, 166
see also coal; gas; oil
Fram (vessel), 15, 77–8
frame analysis, 93–110
and the ACIA (2005), 99–103
and anthropogenic influence, 93, 96–9, 102–4, 106, 108–10
and climate models, 96–7
and climate variability, 93–110
and detection and attribution, 95–110
and framings after 2007, 104–8
and framings until 2007, 100–104
and greenhouse forcing, 97–8
and greenhouse gases, 93–108, 110
and historical records, 95–7
and ICARP (2006), 103
and IPCC reports, 99–104
and ISAC, 99–100, 105–7
and methodology, 99–100
and policymakers, 93–4, 98–9, 102, 108–9
and summary and conclusions, 108–10
and SWIPA, 99–100, 106–8
see also news story that "was" (Arctic climate change as)
France, 5, 56, 79
heat wave (2003), 5
Franz Josef Land, 75–6, 133
frozen saltwater, 157, 168–9
A Functional Glossary of Ice Terminology, 82
future of Arctic, 157–69
from Gaia to the Anthropocene, 164–6
and morality, 160–2
and progress, 166–9
and science and media, 158–60
and social theory and mediation, 162–4

Future Earth, 164

Gaia hypothesis, 164–5, 168
gas, 1, 20, 122–4, 128–9, 131–3, 143–54
Geophysical Research Letters, 40
geopolitics, 1–2, 17, 20, 61, 63–4, 73, 83, 128, 144–5, 148, 162
Germany, 129–30, 146
GHG, *see* greenhouse gases (GHG)
Giddens, Anthony, 167–8
GISS, *see* Goddard Institute for Space Studies (GISS) (NASA)
glaciers, 3, 41, 46, 71, 81–3, 86n3, 104, 138–40, 143, 161
Gladwell, Malcolm, 26, 30–1, 48
global governance, 162, 165–6
"global warming," the term, 6, 29–30
globalization, 1–21, 153, 158, 160–2, 164–8
globalization, climate change, and the media, 1–21
and Arctic region-building, 14–17
and "issue attention cycles," 26
and mediatization, 8
see also mediatization
and meta-event, *see* meta-event
and methodology, 19–21
and natural science pictures, 17–19
and scalar transcendence, 11
see also scalar transcendence
and science and technology as social process, 11–14
and sea-ice minimum as a media event, 1–6
see also sea-ice minimum (Arctic) (2007)
and topical multiplicity, 9–11
see also topical multiplicity
and words, images, and moments, 6–8
Goddard Institute for Space Studies (GISS) (NASA), 45
Gorbachev, Michael, 14
Gordienko, Pavel, 80–1
Gore, Al, 7–8, 30, 100
Gray, John, 167–8

Great Britain, 7, 33, 47–8, 72, 79, 129–32, 138
"greenhouse effect," 7, 26, 29, 33
greenhouse forcing, 97–8
greenhouse gases (GHG), 7, 26, 29, 33, 93–108, 110, 137, 165–6
Greenland, 3, 40, 43–4, 46, 73, 77, 79–80, 87n10, 117–18, 130, 146, 148
"Greenlandic Ice is Melting Faster than Anyone Thought" (2007) (*DN*), 40
The Guardian (U.K.), 19, 27, 33–8, 40–2, 158
pieces published (2003–2006), 35
pieces published (2007–2010), 35
pieces published (2003–2010), 36
sectional breakdown of (2003–2010), 37

Hans Island, 148
Hansen, James, 5, 45, 158
Hasselblad camera, 53
Helland-Hansen, Björn, 75
His Dark Materials (Pullman), x
Holdren, John, 46
Holocene, 71, 84, 164
Hopen Island, 146
Houston, Texas, 1
humanities, 10, 20–1, 158, 162–8
humanity, as planetary change agent, *see* anthropogenic influence
Hurricane Katrina (2005), 5, 30
hurricanes, 5, 30, 70

IAMCR, *see* International Association of Media and Communication Research (IAMCR)
IASC, *see* International Arctic Science Committee (IASC)
ICA, *see* International Communication Association (ICA)
ICARP-II, *see* International Conference on Arctic Research Planning (ICARP-II)
ice-free Arctic Sea, 2, 17, 40–1, 47, 70–86, 88n28, 93, 107, 115, 124, 128, 135, 142, 147, 151, 157

ice-free Arctic Sea – *continued*
  and the Cold War, 79–83
  as "cryo-historical" moment, 71
  and long-term variabilities, 72–3
  and media and the power of
    narrative, 83–5
  and northern expansion of Soviet
    power, 73–7
  and science and politics, 70–86
  and Western responses, 77–8
"ice pad drilling base," 152
ice type/quality, 114, 119–21
icebreakers, 15, 46, 61, 76–80, 138,
  144
ICES, *see* International Council for the
  Exploration of the Sea (ICES)
IGY, *see* International Geophysical
  Year (IGY) (1957–1958)
IMO, *see* International Maritime
  Organization (IMO)
"In a far corner of Greenland, hope is
  fading with the language and
  sea ice" (2010) (*Guardian*), 43
indigenous people, and sea-ice
  change, 18, 94, 114–25
  and coastal erosion, 121
  and hunting, 116–23
  and indigenous knowledge, 114–17
  and local people, 120–5
  and the long-term, 124–5
  and methodology, 115
  and observational data, 117
  and other people, 122–4
  and shipping, 123
  and umiaq, 120
  and use of sea ice and open water,
    117–20
  and walrus, 120–1
  and whaling, 115, 120–2, 130
  see also Inuit; Iñupiat; Yupik
individuals, 3–5, 43–4, 77, 125, 160–7
industrialization, 129
Industrial Era, 164
Inter-American Commission on
  Human Rights, 44, 47
Interdepartmental Bureau on Ice
  Prognosis, 76
interdisciplinary research, 3, 158,
  162–3, 168

Intergovernmental Panel on Climate
  Change (IPCC), 7–8, 29–30, 33,
  93, 98–106, 117, 160–1
  fourth assessment (2007), 7–8, 93,
    101, 103–6, 160
  second assessment (1995), 98
  Special Report on Regional Impacts
    (1997), 101
  third assessment (2001), 33, 100–1
International Arctic Science
  Committee (IASC), 105
International Association of Media
  and Communication Research
  (IAMCR), 164
International Commission for the
  Preparation of the Second IPY
  (1930), 75–6
International Communication
  Association (ICA), 164
International Conference on Arctic
  Research Planning (ICARP-II),
  103
International Council for the
  Exploration of the Sea (ICES),
  74
International Geophysical Year (IGY)
  (1957–1958), 11, 15–16, 55, 59,
  80, 84
International Geosphere-Biosphere
  Programme, 164
International Human Dimensions
  Programme, 163–4
International Maritime Organization
  (IMO), 123
International Oceanographic Courses,
  75
International Polar Year (IPY), 11,
  16–17, 20–1, 59, 61–2, 75–6,
  103, 105, 116–17
  1882–1883, 11
  1932–1933, 11, 76
  2007–2008, 16–17, 59, 103, 105
International Study of Arctic Change
  (ISAC), 100, 105–7
Inuit, 17, 47, 73, 114, 117
Iñupiat, 114, 120
IPCC, *see* Intergovernmental Panel on
  Climate Change (IPCC)
iPhone, 66

IPY, *see* International Polar Year (IPY)
ISAC, *see* International Study of Arctic Change (ISAC)
"issue attention cycles," 26
Izvestiya, 133, 151

Japanese satellite MOS-1, 56
Joint Research and Development Board (JRDB), 81–2
journalism, 159–62
JRDB, *see* Joint Research and Development Board (JRDB)

Kara Sea, 74, 133, 144–5
Kennedy, John F., 4, 6
Kerner, Fritz, 72
King, Martin Luther, 4
Knipowitsch, Nikolai, 74
Kola peninsula, 74–5, 154n1
Komi Republic, 133
Krassin(vessel), 77
Krupansky, Jack, 41
Kyoto Agreement (1997), 30, 161

Landsat system, 55–6
Lenin, Vladimir, 133
liquid natural gas (LNG), 144–6, 154n1
LNG, *see* liquid natural gas (LNG)
Lovelock, James, 165, 168

Makarov, Stepan, 78
Malmgren, Finn, 78
Marginal Ice Zone Experiment (1983 and 1984), 15
Marine Waters and Ice (*Morskievody i l'dy*) (1938) (Schmidt), 76
Maud expedition, 1918 to 1925, 77–8
McCain, John, 46
McKibben, Bill, 158
media
  absence of, *see* news story that "was" (Arctic climate change as)
  and drama, *see* drama, and media
  and globalization, *see* globalization, climate change, and the media
  and mediatization, *see* mediatization

and morality, *see* morality
mediation, 162–4
mediatization, 2–3, 5–6, 8–11, 20, 26–7, 158, 166
  see also scalar transcendence; topical multiplicity
"Meltdown fear as Arctic ice cover falls to record winter low" (2006) (*The Guardian*), 41
"Memos Tell Officials How to Discuss Climate" (2007) (*New York Times*), 42
meta-event, 4, 6–7, 26–7, 48, 49n1, 157
methodology, 19–21, 158
migration, 27, 121
mining, 129–31, 134–43
"Models, Media and Arctic Climate Change" (research project), x
modernity, 9, 154, 167
morality, 159–62, 166
MOS-1 (Japanese satellite), 56
Murmansk, Russia, 1, 14, 144, 154n1
Murmansk speech, 14

Nansen, Fridjof, 15, 75, 77–8
NASA, *see* National Aeronautics and Space Administration (NASA)
Nathorst, A. G., 141
National Aeronautics and Space Administration (NASA), 3, 45, 53, 55, 59–67, 158
  see also Blue Marble photos
National Centre for Atmospheric Research in Colorado, 41
National Geographic News, 61–2
National Snow and Ice Data Center, 2, 21, 41, 54, 83, 85 157–8
NATO, *see* North Atlantic Treaty Organization (NATO)
natural resources, 2, 20, 74, 123, 128–54, 164–6
  see also Arctic energy exploration (1898–1985); fossil fuels; gas; oil
Netherlands, 27, 129, 132
The New York Times (U.S.), 19, 27, 32–3, 35–8, 40, 42, 61, 84, 157
  pieces published (2003–2006), 35
  pieces published (2007–2010), 35

Netherlands – *continued*
pieces published (2003–2010), 36
sectional breakdown of (2003–2010), 37
Newfoundland, 73
news story that "was" (Arctic climate change as), 26–49
and choice of terms, 34–6, 41–3
context of, 27–31
and frames, subframes and topics, 46–7
and ice, water and bears, 38–44
and incidence/contextual frame, 34, 43
and methodology, 32–4
and narrative structure, 34, 40–1, 44, 48–9
and quantitative analysis results, 34–44
and title of story, 34, 37, 40–1
and topic/theme, 34, 38–46
and voice, 34, 45
see also tipping point
newspapers, 2–3, 19, 27–8, 32–8, 45, 144, 159
see Dagens Nyheter (DN) (Sweden); *The Guardian* (U.K.); *The New York Times* (U.S.)
Nikolai Knipowitsch, 75
Nimbus satellites, 55, 58
Nobel Peace Prize, 8, 30, 100
Nobile, Umberto, 77
Nordenskjöld, A. E., 141
Norman Wells, 132
Norse, 73
North American drought (1988), 5, 30
North Atlantic Treaty Organization (NATO), 14, 79
North Pole, 41, 47, 61, 76, 78, 80, 105, 114, 141
"North Pole – 1," 76, 80
North Sea oil, 132
Northeast Passage, *see* Northern Sea Route (Northeast Passage)
Northern Lights (The Golden Compass) (Pullman), x
Northern Sea Route (Northeast Passage), 1, 21, 61, 73–4, 76, 79–80

The Northern Sea Route (Armstrong), 79
The Norwegian Sea (Nansen and Helland-Hansen) (1909), 75
Northwest Passage, 61, 132
Norway, 77, 83, 118, 129–31, 134, 138–40, 152, 154n1
NSIDC, *see* National Snow and Ice Data Center

Obama, Barack, 42, 46, 48
"Obama's revolution on climate change" (2008) (*Guardian*), 42, 48
objectivity, 159–61
oceanography, 15, 19, 75–7, 82
Odeco, 151
Office of Naval Research, 80
oil, 20, 42–3, 47, 85, 122–4, 128–33, 143–54, 158, 166
Okhotsk Sea, 147–8
ozone depletion, 26, 161, 165

Palin, Sarah, 41–2
Panama Canal, 147
Panarctic Oils, 132, 145–6, 150–1
Panel on Arctic Environments, 82
Peary, Robert, 78
permafrost, 17, 82, 86n3, 100–1, 106, 116, 140, 143, 151
Petermann's Geographische Mitteilungen, 72
Pettersson, Otto, 72
Photoshop, 65–6
"a picture is worth a thousand words," 52
see also satellite images of sea-ice minimum
pipelines, 143–6, 151, 154
"planetary boundaries," 165
polar bears, 3, 39, 42–3, 46–7, 114, 122, 125, 140, 158, 161
Polar Eskimos, 43–4, 46
Polar Gas Project consortium (1972), 145
polar map, 61, 63–4, 66–7
polar orbits, 57–8, 64
Polar Record, 79
"Polar Warming," 81

policymakers, 9, 28, 93–4, 98–9, 102, 108–9, 162
Pravda, 144
"progress," 167–8
Prudhoe Bay, Alaska, 131, 143–4
Pullman, Philip, x

quantitative analysis, 34–44

Reuters, 59–60
"Revealed: oil-funded research in Palin's campaign against protection for polar bear" (2008) (*Guardian*), 42
"reverse oil price shock," 152
RICC, *see* Special Report on Regional Impacts (RICC) (1997)
Rio+20 process, 10
Risk Society (Beck), 167
Rossby, Carl-Gustaf, 81
Ruhrgas, 146
Russia, 1, 42–3, 74, 83, 86n7, 117, 129–33, 152
  see also Soviet Union
Russian-American Company, 130
Russill, C., 30

Sadko expedition (1935), 76
"Sarah Palin v the polar bear: who will survive?" (2008) (*Guardian*), 42
Satellite Pour l'Observation de la Terre (SPOT), 56
satellite images of sea-ice minimum, 2, 52–67
  and airplanes, 55
  and Blue Marbles, *see* Blue Marble photos
  and data collection, 57–9
  and "eyes on the world," 54–7
  and interactive map, 61
  and narratives, 53
  and National Snow and Ice Data Center, 2
  and Nimbus satellites, 55, 58
  and orbital position, 57–8
  and photographs, 52–4
  and remote sensing, 52–67
  and Sputnik, 55

"Satellites witness lowest Arctic ice coverage in history" (image) (NASA), 62
Savoonga, Alaska 118, 121–2
scalar transcendence, 10–11, 27, 39, 43, 125, 160–9
Schmidt, Otto, 76
science and technology studies (STS), 12
scientific certainty, 39–40, 45–6, 157, 160
scientific controversy, 7–8, 19–21, 31, 41, 45–6, 49n4, 94–6, 106, 108–10, 160–1, 168
Scott Polar Research Institute in Cambridge, 79
Scottish Spitsbergen Syndicate, 138
Sea Ice Knowledge and Use program (SIKU), 17, 116
sea-ice extent (1979–2007) (figure), 3
sea-ice minimum (Arctic) (2007), 2–6, 17–19, 21n5, 26–7, 59–64, 70–1, 93, 104–8, 114, 121, 125, 157–69
  and absence of reporting, *see* news story that "was" (Arctic climate change as)
  as canary, 102
  and commercialization, *see* commercialization
  and energy resources, *see* natural resources
  and frame analysis, *see* frame analysis
  and future of Arctic, *see* future of Arctic
  and geopolitics, *see* geopolitics
  and globalization, climate change, and media, 1–21
  graph displaying, 3
  and humanity as change agent *see* anthropogenic influence
  and ice-free Arctic Sea, *see* ice-free Arctic Sea
  and indigenous people, *see* indigenous people, and sea-ice change

sea-ice minimum – *continued*
  and media, *see* media; news story that "was" (Arctic climate change as)
  as a moment, 1–6
  and natural science pictures, 17–19
  and the Northeast Passage, *see* Northern Sea Route (Northeast Passage)
  and Pacific sector, 118–19
  and satellite images, *see* satellite images of sea-ice minimum
  and 2012, *see* sea-ice minimum (2012)
  as unique event, 70
sea-ice minimum (2012), 2, 157
sealing, 44, 47, 73, 77, 114, 119–20
SEARCH for DAMOCLES program, 105
Second International Conference on Arctic Research Planning, 100
September, as awareness month, 7, 21, 49n3, 59, 61, 114, 124–5, 157, 159
September 11, 2001, 5
Severnaya Zemlya islands, 76
Shishmaref, Alaska, 118, 121
Siberia, 1, 76, 78, 117, 133, 144–5, 151
siku (seaice), 17
SIKU, *see* Sea Ice Knowledge and Use program (SIKU)
Siple, Paul A., 84, 89n30
Snøhvit oil and gas field in the Barents Sea, 133
Snow, Water, Ice and Permafrost in the Arctic (SWIPA), 17, 100, 106–8, 116
Snow Dragon (vessel), 21
social sciences, 10, 19–21, 49n2, 158, 162–7
Sohio, 131
South Pole, 77–8
sovereignty (Arctic), 2, 15, 77, 105, 134
Soviet Union, 14–15, 55, 63, 71, 73–85, 86n7, 87n12, 88n17, 130–3, 144–8, 151, 153
  and *Glavsevmorput* agency (1932), 73–4
  and power, 73–85
  see also Russia
Special Report on Regional Impacts (RICC) (1997), 101
specialization, 163
Spitsbergen, 75, 129–46, 153–4
Spitsbergen Treaty (1920), 134, 146
SPOT, *see* Satellite Pour l'Observation de la Terre (SPOT)
Sputnik, 55
STI Heritage, 1–2
Stockholm University, 40
Stony Brook Harbor, New York, 115
STS, *see* science and technology studies (STS)
Suez Canal, 1–2
Suomi NPP (satellite), 64
Svalbard Treaty (1920), 73, 77
Svea mine (Swedish), 141
Sverdrup, HaraldUlrik, 76–8, 80, 87n11
Sweden, x, 19, 27, 29, 32–8, 40, 46, 48, 56, 72, 76–8, 81, 86, 87n9, 118, 129–30, 134, 138–42
  see also Dagens Nyheter (*DN*) (Sweden)
Swedish Research Council FORMAS, x
Swedish Satellite Image, 56
Swedish Space Corporation, 56
SWIPA, *see* Snow, Water, Ice and Permafrost in the Arctic (SWIPA)

Telegraph, 59–60, 62–3
television, 2–4, 27–30, 158
terrorism, 27
Thailand, 1–2
thresholds, 26, 48, 55, 122, 165
Time magazine, 84
tipping point, 26, 30–1, 41, 48, 165
The Tipping Point (Gladwell), 26
Tiros (satellite), 55
Titanic sinking (1912), 73
topical multiplicity, 9–11, 27, 39, 116, 160–9
Trans-Alaska Pipeline, 144
Trans-Canada pipeline, 145
Tusaqtuut ("the news season"), 73

United Kingdom, 27–8, 33, 37, 159

United Nations, 5, 10, 14, 38, 98, 100, 109, 146
UN Conference on the Human Environment (1972), 10
UN Framework Convention on Climate Change (UNFCCC), 5, 98, 100, 109
UNFCCC, *see* UN Framework Convention on Climate Change (UNFCCC)
United States, 15, 27–30, 33, 42, 44, 48, 55–6, 63, 79–83, 105, 129, 131, 144, 146, 159–60
US Coast Guard, 144
US Congress, 5
US Department of Defense, 130
US Federal Fish and Wildlife Service, 42–3, 46
US National Academy of Sciences, 80
US National Snow and Ice Data Center, 41, 70
US Navy, 72, 82
University of East Anglia, 159
Urry, John, 167

walrus, 120–1, 124–5
Warsaw Pact, 14
Weber, Max, 166
whaling, 15, 73, 77, 115, 120–2, 130
Wiese, Vladimir, 74–7
Wilhelm, C.L., 141
"wind-chill factor," 89n30
Woodstock, 6
World Climate Research Programme, 16
World Data Centers, 81
"World Scientists Near Consensus on Warming" (*New York Times*) (2007), 41
World War I, 78, 130
World War II, 14–16, 76, 79–80, 132

xenophobia, 27

Yermak icebreaker, 78
Yupik, 114, 117, 121

Zeppelin expedition (1928), 77
Zubov, Nikolay, 74–7, 79, 87n12

Printed and bound by CPI Group (UK) Ltd, Croydon, CR0 4YY